Praise for *Carbon, Capital, and*

"The energy transition era requires oil and gas companies to reframe how they fulfill their role and responsibility in our world. Companies will need to leverage digital to transform their operating cost structures, human productivity, environmental surveillance, compliance to regulations, and safety standards to ensure sustainability. How they go about this journey calls out for guidance.

Carbon, Capital, and the Cloud is the industry playbook to equip companies with insightful and concise direction on how to be digital to garner impactful change, not just to check a box. This is an eye-opening read that delivers a swift kick to the rear of any executive looking to avoid their company being left behind. Highly recommend this book."

ROB FIORENTINO, P.Eng., VP of Operations and EH&S, Crescent Point Energy

"Safety and reliability in the oil and gas industry were reasons to avoid or slow-play digital transformation. The pandemic turned this assumption on its head. Now, can digital transformation help companies achieve cultural change and respond to decarbonization while still being profitable? *Carbon, Capital, and the Cloud* weaves a practical narrative with examples across the value chain.

I highly recommend Geoffrey and Ryan Cann's reference-worthy material. Pro tip: the interview guide is a great starting place for energy execs to ask the right questions."

HEATHER CHRISTIE-BURNS, vice president, transmission pipeline unit, Pembina Pipeline Corporation

"Oil, gas, and energy companies pursue a bold vision to deliver safe, reliable, and sustainable energy products and services focused on the customer and enabled by innovation. *Carbon, Capital, and the Cloud* is a playbook for oil, gas, and energy companies as they streamline their transition to sustainable energy in the cloud."

BENJAMIN BEBERNESS, global energy industry leader, SAP

"Having worked in and with digital oilfield challenges since the early days and experienced both success and failure, Geoffrey has once again masterfully aligned the digital benefits message for the individual worker and the corporate business."

DAVID SMETHURST, executive, entrepreneur, thought leader—energy and technology

"Geoffrey Cann's next installment in his series on how to digitize the complex world of oil and gas as it redefines itself to energy is captivating. Addressing the need for change is not just about the increased use of digital tools within traditionally manual processes, but also about the reengineering of an engineering-led operational construct built on a digital core.

This is a must-read not just for the oil and gas engineer, executive, and team member, but for anyone who sees the opportunity to reengineer the energy industry from the ground up."

NISH KOTECHA, chairman, Finboot

"Once again, Geoffrey Cann takes his readers on a swift journey through energy's extraordinary digital future. To avoid being left behind by the digital transformation and to thrive on this uncertain journey, energy executives everywhere need to put *Carbon, Capital, and the Cloud* on their must-read list."

DAVID FORSBERG, managing partner, Ascent Energy Ventures

"*Carbon, Capital, and the Cloud* is so timely and pertinent for oil and gas and other energy companies focused on digital transformation and energy transition. It covers the full range of important themes that are being discussed daily within big energy companies. It is a fantastic resource that captures the richness of these discussions in one easy-to-read place and will serve as a thoughtful playbook for companies seeking to accelerate into the future in this world of energy system change. Highly recommended."

SUNGENE RYANG, executive director and CEO, GS Beyond

"Geoffrey and Ryan have delivered a playbook that will serve as a valuable reference as oil and gas companies journey through the many distinct phases of digital transformation. The guidance provided breaks down the real challenges that organizations will face and the strategies to navigate these opportunities successfully. This book is a must-read for energy executives and board members alike."

CAMERON BARRETT, CEO Field Safe Solutions

"*Carbon, Capital, and the Cloud* is more than a sequel to *Bits, Bytes, and Barrels*. It exposes the powerful impact of the trifecta at the core of a brave new economic model—energy transition, decarbonization, and digitalization—and uncovers the trends and critical signposts for every industry leader, manager, and investor. The authors lay out a roadmap to the digital energy model, where oil demand reaches an apex, capital allocation is automated, and daily decisions in oil and gas are based on real-time data gathered by sensors and interpreted by machine algorithms. This is a survival guide for every industry executive in the sector developing a strategy to thrive by surfing the next wave of the digital tsunami."

EDUARDO RODRIGUEZ, president and director,
FlatStone Capital Advisors Inc.

"People need to really understand the immense opportunity that awaits the oil and gas industry if a digital future is fully embraced— Geoffrey Cann does a superb job breaking this down for the reader. His scientific and well-researched approach in making a case for how the energy transition can, and should, capitalize on digital assets is groundbreaking. He is one of the few thought leaders in this space, and I will continue to utilize this book in the classroom."

ANN BLUNTZER, PhD, associate professor, TCU Energy Institute

"I am so fortunate to have discovered Geoffrey Cann's witty yet honest perspective at a major inflection point in my thirty-plus-year career in the oil and gas sector. As a follow-up to the informative *Bits, Bites, and Barrels*, released in 2019, this playbook arrives with near perfect timing, just as the promised digital transformation is finally beginning to take hold in our industry. Geoffrey and Ryan Cann artfully leverage decades of thought leadership experience and practical-use case research evidence to guide us step-by-step through this critically necessary conversation. Their growth mindset adjusts for uncertainty while providing needed clarity by educating us about what we can do to create a more sustainable future for everyone. As a consultant seeking to influence adoption of digital innovations, I personally found the included glossary of industry and technical terms used in context an invaluable resource, informing and connecting the reader to the complex dynamics now shaping the energy transition."

RANDELL MCNAIR, DBA, information asset management consultant

"The global oil and gas industry is vitally important to our day-to-day lives, but we rarely give it a thought. Now, the pressure is on to transition to new energies, and to decarbonize all existing products as quickly as possible. Hidden in this call to action is the undeniable fact that our world will be dependent on fossil fuels and its spinoff products for decades to come.

Thoughtfully written, deeply practical, and laser-focused on the things that truly matter, *Carbon, Capital, and the Cloud* sets out how all companies in the oil and gas value chain can secure their futures by accelerating their adoption of digital innovation. I particularly enjoyed the case studies that prove the point. Well worth the investment."

PAUL KURCHINA, SAP community evangelist

CARBON, CAPITAL, AND THE CLOUD

GEOFFREY CANN AND RYAN CANN

CARBON,

CAPITAL,

MadCann
PRESS

A Playbook for
Digital Oil and Gas

AND THE CLOUD

We dedicate this book to

Marjorie, for her patience and support as we turned a motley collection of disjointed blog posts into a proper book.

Meaghan, for too many reasons to list.

Cataloguing in publication information is available from Library and Archives Canada.
ISBN 978-1-77458-223-7 (paperback)
ISBN 978-1-77458-224-4 (ebook)

Page Two
pagetwo.com

Edited by Lesley Erickson
Copyedited by Rachel Ironstone
Cover design and illustration by Cameron McKague
Interior design by Taysia Louie
Interior illustrations by Cameron McKague
and Taysia Louie

geoffreycann.com

"Oil companies are generally built on the same set of assumptions. Operations first, everything else is a cost to be minimized. Only operations skills matter, and the future will be the same as the present."

VICE PRESIDENT OF AN OIL AND GAS PRODUCER

CONTENTS

INTRODUCTION

THE DEBATE around whether digital innovation will impact the oil and gas industry is well and truly over. First, capital market pressures have built to a point where the industry has needed to respond. Second, the pandemic of 2020 forced the industry to deploy some digitally enabled changes to its business model to survive. These two forces have combined to create a kind of Great Leap Forward. One industry executive described the scale of the resulting shift as the equivalent of accomplishing twenty years of change in just twenty months.

Let's begin with capital markets, and the pressure that has been increasing there for some time. After a hundred years, Exxon was dropped from the Dow Jones Industrial Average index, and institutional shareholders voted three dissident directors to its board.[1] Funds such as BlackRock announced that investment in the industry would now depend on clear and accountable actions to address climate issues.[2] Norway's sovereign wealth fund announced full divestiture of fossil fuels from its portfolio.[3] Royal Dutch Shell lost an important court case in its home market because its efforts to decarbonize were too slow.[4] Pipeline companies in North America have abandoned projects that otherwise serve to improve the energy security of the continent.[5] The International Energy Agency (IEA) published studies showing that renewable energies are now cost competitive in all markets relative to fossil fuels, and that all fossil fuel investments must halt if the strictures of the Paris Climate Accords are to be met.[6]

The pandemic forced a swift response. Social distancing measures such as working from home were quickly put in place, leveraging the latent capability to **telework**, which was part of the technology infrastructure deployed previously by the industry. Managers rearranged paper-based office work that formerly depended on people working in physically close teams and accelerated the adoption of electronic documents and workflows. Even trading teams were banished from their purpose-built trading floors and sent home to carry on from kitchen counters, second bedrooms, and garages. Closed borders blocked the usual level of travel for training, conferences, trade shows, deal making, and research, all of which were quickly reconfigured for a virtual world. Slow-moving projects to create remote control rooms and digital oil fields hustled over the finish line and into production.

Quietly, oil and gas companies are reporting the surprising benefits from the shift. Permanent cost reductions are dropping into place. The productivity of the workforce has maintained its usual level, and in many cases is increasing. Some assets, such as real estate, have suddenly become surplus to need. Emissions levels are improving, incidentally from more efficient work design that entails less driving, and explicitly from better decision-making on investments that involve emissions in some fashion. Capital is still tight, but mergers still take place.

The industry has received a number of critical messages from this period in its history:

- Capital did not return to the industry even during the pandemic, and governments are more resolute than ever to promote alternative energy systems.

- The oil and gas industry can successfully implement change much more quickly than was assumed. The abundance of caution that typifies oil and gas decision-making may be overly strict.

- Oil and gas operations, previously dismissive of digital innovation, and where most operating cost and the bulk of the capital expense resides, also benefit from digital improvements.

- The scale of benefits unlocked by adopting digital innovations, in cost reduction and productivity gains, can greatly exceed expectations.

- Digital innovations play a surprising role in both enabling business improvement gains and improving environmental performance.

- New business models are accessible once the basics of digital innovations are in place.

The survivors will be those who figure out how to take these insights on board, embed them in their business strategies, and move decisively to reposition their operations along digital lines.

The How of "Digital"

Even prior to the events of 2020, we were asked about when we would turn to the next installment of this epic story of change. Oil and gas professionals were already sensing that digital was an inevitable feature of the future in oil and gas, and some companies had already moved ahead of their peers in embracing this future. As Cory Bergh, vice president of financial and information services at NAL Resources, put it, "We have to do things we never planned to do, and we don't have the skills to do it, and we have to do it faster."

Now that the whole industry has accepted the *why* of "digital," the next question is *how*: how to do digital better, faster, and with the greatest impact.

This book sets out to answer these questions:

- What are the drivers of change for oil and gas, and how can they be best harnessed?

- Which digital technologies are proving themselves to be instrumental to the industry, and how are they being deployed?

- How are business models evolving to take advantage of digital change?

- How is the talented workforce of oil and gas professionals best engaged in the change?

- How are companies successfully implementing digital change?

It is unnecessary to read our previous book, *Bits, Bytes, and Barrels*, about digital innovation in oil and gas to benefit from this book, but readers might find it helpful.

Who Should Read This Book

Our overarching goal with this book is to help the industry accelerate its adoption of digital innovation. We are currently deeply dependent on fossil fuel products, and we need to accelerate our transition to a different fuel mix. The industry now agrees that digital innovations are a pathway to lower costs, better productivity, and reduced emissions. The COVID-19 pandemic proved that the industry can change quickly. The question is how—how to move faster.

Aiming primarily at the oil and gas industry, we set out to write this book with several goals in mind, namely, to help interested parties do the following.

Focus attention: The technology landscape is complex and evolving, and almost all innovative technologies appear to offer significant benefits. Those that are unlocking new business models warrant greater urgency.

Embrace the whole industry: The oil and gas industry is broad—onshore, offshore, **upstream, midstream, downstream,** refining, distribution, and retail. Digital innovation impacts the breadth of the industry.

Guide leaders: Those tasked with driving digital change are looking for specific tactics, insights, and moves they can make. They are particularly challenged with helping oil and gas workers, who have never experienced a combined energy transition and a digital transformation, cope with feelings of confusion, abandonment, and uncertainty.

Speak plainly: The digital industry and the oil and gas industry have their respective terminology, abbreviations, and insider language. To be effective, the conversation needs to cut through the obscurity and provide meaningful insight. That said, the use of some industry terminology is necessary, and to aid understanding, there is a glossary provided at the back of this book. The first time a term appears, it is set in **bold** to signal its inclusion in the glossary.

With this in mind, the book will appeal to many different audiences:

- oil and gas industry leaders preparing their companies for an uncertain future;
- industry professionals navigating their careers through these sweeping changes;
- students in petroleum engineering and geology looking to better understand the evolving industry;
- entrepreneurs wishing to disrupt the industry through new business concepts;
- supply chain companies (field services, hardware, technical goods) addressing digital in their businesses and with their customers; and
- technology companies wishing to penetrate the industry and seeking points of entry.

Oil and gas is a gigantic industry, reaching into every country on the planet. Regardless of what aspect of the value chain (upstream, midstream, downstream) is of interest, you will likely find this book of value. It is weighted to the asset-centric, business-to-business parts of the industry.

Navigating the Content

Depending on your areas of focus and interest, you may find some chapters more valuable than others. What follows is a brief overview of the content by chapter, so that you may dip in and out as best suits your needs.

Chapter 1 sets out the forces of change facing the industry, which include the ongoing impacts of the COVID-19 pandemic and pandemics to follow, **capital** market demands for improved resilience, and environmental pressures to reduce the ancillary impacts of the industry. It also looks at the narrative of the industry and the debate of the future of oil and gas. It should be of interest to all readers.

Chapter 2 reviews leading digital innovations and how they are being used in the industry. This chapter will be of particular interest to technology companies and entrepreneurs seeking to understand how receptive players in the industry have been to digital innovations therein.

Chapter 3 discusses the new **business model**s in oil and gas that are creating fresh opportunities for entrepreneurship and a cleaner and safer future. All readers should find this chapter inspiring, as these new business models have the potential to disrupt the industry at scale.

Chapter 4 reveals the impacts that these changes are having on people and talent, since digital change is really about change management. Anyone in a role that includes leading teams or helping people come to grips with digitally driven change will relish this chapter.

Chapter 5 presents nine case study companies from across the industry (equipment suppliers, upstream, midstream, downstream, integrated oil and gas, and services) that are investing in driving change, and reveals the tactics they employ and the results they achieve. Those tasked with leading or driving digital change in the industry will find inspiration in their work and the collection of tactics very helpful.

When you reach the end of this book, you will better understand the changes that the industry must embrace for its success, and you will have a solid grasp of the successful tactics that leaders need to deploy to accelerate the transition.

You may have noted that this book has two authors; for simplicity's sake, we have written from a first-person perspective. For clarity, all of the anecdotes reflect Geoffrey's direct experience from working in the industry.

Notes

1   Pippa Stevens, "Exxon Mobil Replaced by a Software Stock after 92 Years in the
    Dow Is a 'Sign of the Times,'" CNBC, August 25, 2020, cnbc.com/2020/08/25/
    exxon-mobil-replaced-by-a-software-stock-after-92-years-in-the-dow-is-a-sign-
    of-the-times.html; "ExxonMobil Announces Final Results in Election of Directors,"
    ExxonMobil, June 21, 2021, corporate.exxonmobil.com:443/News/Newsroom/
    News-releases/2021/0621_ExxonMobil-announces-final-results-in-election-of-
    directors.
2   Alex Kimani, "BlackRock Is Turning Up the Heat on Oil Companies," OilPrice.com,
    January 27, 2021, oilprice.com/Energy/Energy-General/BlackRock-Is-Turning-Up-
    The-Heat-On-Oil-Companies.html.
3   Lars Erik Taraldsen and Ott Ummelas, "Norway's Sovereign Wealth Fund
    Scrutinizes Its Oil Holdings," World Oil, February 4, 2021, worldoil.com/
    news/2021/2/4/norway-s-sovereign-wealth-fund-scrutinizes-its-oil-holdings.
4   David Vetter, "'Monumental Victory': Shell Oil Ordered
    to Limit Emissions in Historic Climate Court Case," *Forbes*,
    May 26, 2021, forbes.com/sites/davidrvetter/2021/05/26/
    shell-oil-verdict-could-trigger-a-wave-of-climate-litigation-against-big-polluters.
5   Joe Walsh, "Keystone XL Pipeline Project Abandoned after Biden Yanks
    Permits," *Forbes*, June 9, 2021, forbes.com/sites/joewalsh/2021/06/09/
    keystone-xl-pipeline-project-abandoned-after-biden-yanks-permits.
6   IEA, "Renewables 2020: Analysis and Forecast to 2025" (Paris, 2020), iea.org/
    reports/renewables-2020.

1

THE FORCES
OF CHANGE

"There are two mega trends that are affecting our industry. One is energy transition; the other is 'digital.' Everything that we do from a change point of view connects to those two strategic drivers."

DR. JOHN PILLAY, SVP Transformation, Worley

I N LATE 2019, I was a guest on a podcast to discuss the current state of digital adoption in the oil and gas industry. Digital adoption appeared to be racing forward under full power. Earlier in 2017, the IEA had published a major study outlining the impacts that "digital" would have on the fortunes of the sector. That same year, the highly influential CERAWeek event declared digital to be the latest must-do in the industry. The leading digital companies had started their domination of capital markets.

The interviewee posed the metaphorical question: If digital adoption were a baseball game for this industry, what inning would we be in?

We're in the early innings, I declared with considerable optimism. The game was clearly underway, a handful of digital technology companies that focused exclusively on the industry had emerged, and there were some modest successes.

In hindsight, I was completely wrong. The game had barely started. In the weeks that followed, BP released its latest forecast of supply and demand, predicting a fast-approaching demand peak. Capital market pressures, environmental and social changes, and the pandemic acted in concert to decisively propel digital investments throughout the industry.

Some of these investments yield significant benefits across multiple dimensions at once—environment and social, cost, productivity, growth and capital optimization—thereby accelerating digital transformation. I call this the digital sweet spot.

The question is no longer "Is digital a good idea?" but "How do we accelerate our adoption of digital?"

THE DIGITAL SWEET SPOT

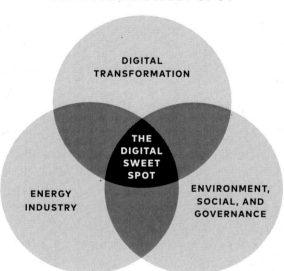

Demand Uncertainty

The major sweeping change that is recasting the industry is the shifting demand profile. In a startling announcement at its annual general meeting in 2020, BP called peak oil demand a reality and indicated that it would arrive far earlier than previous modeling efforts had predicted.[1]

Their modeling across a handful of likely scenarios of the world of energy pointed to flat and declining global demand for oil, which prompted BP to embark on a fresh strategy to reduce its exposure to oil. Earlier in 2017, Shell announced a similar plan (a reorientation of resources to new energy) but without the hammer statement of peak demand.[2]

Almost two decades ago, BP rebranded itself with its new sunflower logo and catchy handle, "beyond petroleum," which translated into exactly nothing strategically. There is a risk that BP is again ahead of itself; that we're far from peak demand, and that the world will soon go on as it has. Transportation demand, which is based on 1.2 billion combustion engine cars, 300 million heavy trucks, fifty thousand military and commercial ships, and thirty thousand airplanes, will change only slowly.[3]

Almost all products (except perhaps the air we breathe and the water we drink) go through a demand S-curve, where product demand slips into decline after a peak. But oil is not the same as CD players, flip phones, or boom boxes. Petroleum is deeply entrenched in our way of life.

The Industry's Public Response

Many oil companies, suppliers to the industry, and nation states dependent on oil have mobilized consulting studies and board meetings to discuss BP's position. Responses will likely vary, typically to justify a do-nothing action plan:

- Some companies will studiously ignore BP until they can formulate their own statements to manage their investors, employees, and communities.

- Some will dismiss the research as just "one man's opinion" and highlight all the tricky assumptions in the analysis that could yield a different answer.

- Some will suggest the analysis is self-serving. They'll point out BP's track record of climate and change missteps, its ill-timed rebranding twenty years ago, its much-publicized problems in Russia and the Gulf of Mexico, and its new-ish management need to break with the past.

- Some will pretend everything is fine, arguing that the huge structural demand that drives the sector is hard to change and will be enduring.

- Some will count on the short-term news cycle to push the story down the reading stack in the hopes that it will simply disappear.

- Some will be heartened by the post-pandemic energy shortages that point to swift demand recovery, evidence that society wants to return to business as usual.

- Some will be content to play wait-and-see, thinking it's too soon to draw any firm conclusions. (Of course, auto executives said the same thing about Tesla a decade ago, and retailers everywhere ignored Amazon too.)

Regardless of their stance, all companies in the industry are now at least alert to the coming changes in oil demand, the impacts and timing of which are still unknown, but predicted.

I'VE BEEN through many oil market cycles, and there is but one truth. Only the low cost survive. My first downturn was in 1985 when I was working for an oil company in Toronto. The demand for oil at the time was around forty million **barrel**s per day (mmbpd) but the available supply was around sixty mmbpd, most of which was from the Middle East. Since all that oil wanted to be consumed, the market collapsed, and prices fell from $30 to $10 **USD**.

To rid itself of surplus staff, the company launched a staff financial buyout scheme called Career Change Assistance Program, or C-CAP. Us staffers morbidly called it "Kneecap" or "Decap." I was newly married and in two weeks was supposed to transfer to the upstream unit in Calgary, but my transfer fell apart.

These cycles can get very personal very quickly. And it takes close to a decade for prices to recover. The company survived because it moved decisively to keep its costs low and its debt restrained.

Ninety Days of Extremes

Each cyclic downturn in oil is unprecedented, but each cyclic downturn is unprecedented in its own particular way.

For example, the price spike of 1973 was due in large part to the shortage of oil tankers needed to move crude to market (shipping costs eventually collapsed with the arrival of new carriers). In Canada, the industry has faced distressingly low prices since 2014 because of a lack of pipelines. The shocks to supply, demand, or both, rarely have the exact same cause every time.

All these downturns have similar consequences. Prices collapse, many players go out of business, layoffs are widespread, and it takes a few years (often a decade or more) for the industry to recover.

I have personally survived market routs in 2001, 2008, 2014, and 2020. In 2020, the most recent market upheaval, the cycle was about abrupt changes in both supply and demand.

Disagreements over Supply

To understand the supply shift, we need to step back to 2008. According to BP, North American oil supply grew from 13.1 mmbpd to 22.6 mmbpd from 2008 to 2018.[4] This much growth this quick was unprecedented. Russian production overall barely budged, 12.7 to 14.5, and the suppliers from the Middle East only grew from 26.5 to 31.7. North American oil production growth was greater than that of the Middle East and Russia combined, and the oil was flowing to Asia, where the need for oil had grown from twenty-six mmbpd to thirty-six mmbpd in a decade.

American shale oil is relatively more expensive, technically complex to produce, and financed by borrowings on capital markets. It has taken market share from Middle Eastern and Russian oil resources, which are less costly and more easily produced by national champions. This might make sense if that expensive American oil commanded a higher price because it was somehow qualitatively better, but it's not.

By February of 2020, the economic ravages of the pandemic and its impact on oil demand were starting to emerge. China is both the biggest buyer of oil globally and the starting point for the pandemic. Both the customer (China) and the supplier (OPEC) could see the impacts of China's lockdown before anyone else. OPEC wanted deeper cuts, but Russia balked, and the producers decided to flood the market with oil.

Any eventual cuts agreed to by the producers are based on recent sales volumes, so the key to minimize the impacts of cuts is to maximize sales. Whether the oversupply was poor decision-making or a shrewd move to push the US out of the market doesn't matter. Supply surged ahead of demand by twenty mmbpd over March and April, leading to heaving inventories of 1.2 billion barrels, straining in storage.

In addition, the North American oil industry was in some pain. The **fracking** segment was already under siege because capital markets were growing increasingly suspicious about the fracking business model and the mountain of debt that these companies had taken on board. These are not low-cost businesses, and many would not survive.

Where supply meets demand is price, and we've seen the effects. The price of oil fell 85 percent within the year, from $70.25 to as low as $9.12, and in some markets under some conditions, the forward price was negative. The quantum of the fall is not without precedent— it fell further in 2008, from $139 to $45.[5]

The pain in the industry has not been uniformly shared. The control of 85 percent of the world's reserves is by governments that are highly dependent on the proceeds from oil sales and are relatively unresponsive to price moves, **carbon** pressures, activists, and capital markets.[6] They are trimming to weather through, but they're not necessarily changing course.

Managing the Fallout

In the commodity oil industry, the price of the product is set by the marginal barrel. That is, in theory, when all demand but one barrel is fully supplied by the market, the buyer of that last barrel will pay a little more to get it, and in doing so, lift all prices at least a little.

Similarly, the marginal barrel sets the price when the market is oversupplied. In 2014, as the US was ramping up its shale plays, the market swung into oversupply by a mere 1 to 2 percent (or between one and two million barrels per day).[7] That oil was conveniently tucked away in storage, but once storage was full, the holder of the marginal barrel with no customer practically gave it away, causing short prices to collapse for everyone.

This played out again with stark consequences at the beginning of the pandemic in February 2020 when Saudi Arabia and Russia both decided to flood the market with their oil in the face of the demand collapse caused by the pandemic.[8] Prices promptly fell, and in some markets under some circumstances, prices went negative altogether with suppliers paying customers to take the product.[9]

If BP is largely correct, we are now, globally and much faster than anticipated, heading into a world that is structurally overbuilt for the demand, with the price of the product to be set by the marginal barrel.

The problem is that oil-producing states, provinces, territories, and nations cannot deliver a managed market structurally and are individually overly dependent on the riches thrown off by the most profitable global commodity that has been in perpetual growth mode for over a hundred years (a very long S-curve). There is no incentive, and in fact, substantial disincentive, to try to manage supply to the demand.

The Consequences

The world has little experience in dealing with a permanently flat or declining oil market and no experience at all in managing an energy transition, which is playing out in late 2021. Oil and gas prices are rising very sharply as economies accelerate out of the pandemic following two years of low investment in energy infrastructure. Here are just some of the consequences that may well follow.

Structural advantage will favor the low-cost producers. That might not be OPEC, whose members are highly dependent on high oil prices to balance their national treasuries and pay for their social programs. There is space globally in the market for low-cost players everywhere. Upstream portfolios will be quickly reconfigured, leading to a wave of transactions and portfolio shaping—along the same lines as what happened in the gas industry.

At some level, capital markets price BP's negative view of the energy industry into their expectations for share value growth and future dividends. This has the effect of depressing all oil company stock prices. Capital for growth and expansion will be harder to secure for those producers outside of the national oil companies, and for those without a compelling story about cost leadership.

Oil companies will high-grade their projects, and only the very best—with the fastest time to market, the lowest capital cost, the highest reliability, the lowest operating cost, the least impact on the environment, and the highest possible production—will be sanctioned. Everything else will be at risk of being mothballed or stranded as boards see that shareholders are protected.

Existing production is now on notice to prepare to be shut down, sold off, or run out, unless there is a plausible plan and meaningful measurable action to become the low-cost asset. This applies to basins, wells, and facilities throughout the entire value chain. Even gas stations will need to transform or consolidate. Managers everywhere should take this to mean "we are now on war footing" with our costs and productivity.

Traditional suppliers to the industry (engineering firms, equipment companies, consultants) whose services reflect the commodity price will need to find new sources of revenue as capital projects vanish and cost pressures everywhere else pinch margins. This should trigger a wave of fresh innovative thinking about the industry, galvanized by successful changes introduced during the pandemic.

Employees in the industry will rethink their career choices. Those who have left the industry will feel vindicated. Long-service workers will be content to just wait it out, but their stock options and pension exposure to the industry will call for personal retirement fund diversification.

The biggest worry is that the talent pipeline influencers (parents, school counselors, social media, youthful activists) will more firmly steer youth away from the sector. The fallout will show up in five short years. Leaders will need to redouble their efforts to keep their people in the game, and they'll need a much more positive story to tell.

Rentier state governments will urgently need to fundamentally rethink their economies as revenues from the industry experience a double fall (reduced volumes and lower royalties because of the lower price). Lower tax revenues mean weaker schools, poor health care, and imperiled retirements.

Short countries (that is, importers of oil) will revisit with renewed interest any green plans. Continued investments in oil import infrastructure will increasingly look like stranded money.

The biggest winners in the shakeout will be those with innovations that work to take out cost. The pandemic has shown the industry how to change fast, and many companies will need to do exactly that.

Digital Lag?

If you are raising money in capital markets for an oil and gas venture, no matter the segment of the industry, chances are you will be asked how you are dealing with technology-driven change. The implicit message is that capital markets suspect that the industry is not fully prepared to embrace digital innovations.

Is Oil and Gas Already Digital?

Capital markets ask about technology-driven change because so many other sectors failed to anticipate the impacts of digital innovations and delayed or avoided acting in the face of the threat. This do-nothing strategy resulted in the permanent destruction of capital. Retail has seen thousands of store closings because of Amazon. Video rental outlets vanished because of streaming. Newspapers have folded because ad revenue has swung decisively to Facebook and Google. Ubiquitous online communication **apps** have taken the market away from telecom providers and phones. Even maps are being displaced by widespread **GPS**. Sector after sector has been transformed.

Oil and gas professionals are quick to declare their industry highly "digital" already. In some very specific respects, it is a fair conclusion to reach. The upstream sector in particular is very data intense and has been collecting and interpreting gigantic **seismic** datasets for decades now. Big oil companies historically match governments in their ownership of super-large data centers and data-processing supercomputers. The industry has a history of selling its **subsurface data** at a level distinct from other sectors (save other resources industries). The sector connects almost all of its infrastructure to **SCADA** (supervisory control and data acquisition) equipment.

Digital Tourism Is Over

Virtually every large oil and gas company is now "doing something" in digital. It's clear that the boards and management now get it. A few have put **digital native**s on their boards. Search the org chart, and you can probably identify the digital innovation team that is toiling away on digital, such as a small group that is trying to shepherd a portfolio of small-scale initiatives to scout out candidate solutions, launch some trials, test some ideas. Some companies even have investment funds that take positions in promising new business ventures.

If anything, it's the supply chain to oil and gas (the thousands of specialized companies that sell services, supplies, and equipment to the industry) that are not very digital.

But the oil and gas approach to digital is consistently out of sync with a digital world that is expanding in capability on an exponential basis every eighteen months. The oil and gas industry business model is predicated on the scarcity of data, high-cost storage, expensive computational horsepower, a few large business partners, and a stately pace of change. Our overall business reality is rapidly shifting to a super-abundance of data, unlimited storage, flexible on-demand analytics, lots of clever startup innovators, and a hyperkinetic business cadence. A single digital company, Amazon, designs, tests, and introduces over fifty million software updates a year, during which time an oil and gas company might launch a few hundred.[10]

To describe the industry as "already digital" is a mischievous misdirection, if not categorically false. To quote Customer Centric and Optimization Technology Director Tomas Malango, from Repsol, "We need to improve our digital skills as a 'sixth sense,' to become better professionals."

The Problems of Perception

It is worrisome that some industry insiders think it helpful to paint the industry as on par with digital industry leaders.

It communicates to next-generation talent that there's no opportunity in the industry. This isn't healthy. The industry already faces strong headwinds from society, and the youth contemplating their career options have ample opportunities elsewhere. Oil and gas should instead be presenting itself as the center of digital innovation among heavy industries.

Capital market participants need strong encouragement to invest in oil and gas. There are a number of movements around the world to divest or severely restrict capital investment in the fossil fuel sector.[11] The industry needs a new narrative; it is not just leveraging technology but also pioneering cutting-edge technologies.

The "already digital enough" mindset robs the industry of much-needed attention from digital entrepreneurs. Since digital is impacting all industries at the same time, just not at the same pace and intensity, entrepreneurs and startups have plenty of opportunity to chase. Without clear signals of interest from the industry, these entrepreneurs will direct their attention elsewhere.

A Late Adopter

There's ample evidence to suggest that oil and gas is a latecomer to the wave of digital change.

Digital Vortex, the book cowritten by IMD and Cisco, looks at a dozen industry sectors to forecast the timing and impact of digital innovation.[12] They conclude that oil and gas is eleventh out of twelve sectors, well behind retail, banking, transportation, and technology. Only the pharmaceutical industry will feel the effects of digital later. I reach similar conclusions in my book *Bits, Bytes, and Barrels*, in which I look exclusively at the various segments of the oil and gas industry (upstream, midstream, downstream, services).[13] The IEA, in their global study about the impact of digital on energy, argue that oil and gas adopts digital features relatively slowly.[14]

This makes complete sense. It's much easier and cheaper to design digital into a new asset than to add it after the fact to an older one, especially in a world of heavy steel assets. More than 85 percent of the assets in oil and gas were in place before digital was even a thing.

The window for adding digital smarts to an operating asset is usually pretty tight—confined to the regular **turnaround**, which is at most only a few days in duration. For **offshore** platforms, the window is even tighter, particularly if hardware is involved.

Oil and gas is a cautious industry and risk averse. For this, society is grateful. Spills of crude oil are hard to clean up, gas pipeline ruptures turn into fireballs, and the product is toxic. The industry applies the highest standards for safety and environmental compliance. Equipment must meet exacting engineering standards, and the introduction of new equipment, including digital, is done under strict management of change (**MOC**) processes.

Economically, it hasn't made much sense for the industry to invest in digital technologies. The downturn in commodity prices in 2014 scaled back the industry's ability to invest, and the limited capital available tends to go towards growth in reserves or production. With conventional oil basins operating only for a few more years, it is hard to justify investing in new technology. It's challenging for managers to demonstrate with clarity that a digital investment will pay off.

The culture of the oil and gas industry is also biased against external digital innovation. The industry relies on incumbent industry insiders to design and introduce new solutions. Those insiders are often tied to legacy technology architectures and designs that date back decades and were never built for the open environment within which digital thrives. They are incentivized by contracting models and licensing structures to keep small firms at bay. Big Oil struggles to figure out how to work with small, nimble startups.

Oil and gas is a long way from being digital. There is still untold opportunity for young people and tech entrepreneurs to find fortunes in the industry.

Just ignore the message that the industry is a digital leader. It's not.

Environment, Social, and Governance Pressures

A second question posed by capital markets is about company responses to environment, social, and **governance** pressures, or **ESG**.

Many people wonder what ESG means. The concept is that companies will take ESG factors into account in their decision-making, so that they consider the long-term consequences of their choices.

I'm simplifying, but environmental considerations can include the impacts of enterprise on water, air, soil, land, plants, animals, fish, and oceans. These can be cumulative and absolute impacts that affect the overall balance of nature, the ability of natural systems to repair and rejuvenate, recover and renew. Social considerations can include impacts on communities, Indigenous populations, urban and rural settings, the disadvantaged, and the developing. Governance can include how decisions are made, and whose views are considered, including considerations such as labor groups, Indigenous peoples, communities, political systems, regulators, and capital markets. Again, impacts can be both cumulative and absolute.

ESG thinking has come about because, traditionally, the production and consumption decisions that countries, enterprises, and individuals make typically place a priority on short-term and narrow criteria, such as achieving the lowest price, maximizing shareholder needs, meeting capital market targets, or satisfying regulators, not on satisfying these broader, longer-stride societal factors.

We Do Not Have a "Planet B"

It might seem absurdly obvious, but our planet is our only home. We all have a stake in it, whether we are from rich countries or underdeveloped nations, regardless of our background or age. Our planet is our collective responsibility, and we're accountable to not just part of the Earth, but all of it, from its plants and animals to its deserts and oceans, from its air and water to its farms and factories. We do not have an alternative planet to move to, although I am appreciative of the billionaires keen to colonize Golgafrincham.[15] Earth is indeed a home, because it's where we grow up, eat our meals, raise our kids, celebrate our successes and failures.

In this light, we should make decisions that help us keep the whole of the planet intact, safe, and long-lasting. All things feature a steady level of degradation: My neighbor's roof leaks, and they need to replace it. But we can't replace the planet, so what we need to do instead is keep it from degrading to a point at which it can no longer be our home.

Ironically, Earth is structurally unsafe for us. There's a lot of visible danger on our planet, from wild storms, to searing heat, to destructive fires, to droughts, floods, volcanoes, earthquakes, and tsunamis. We are routinely under threat, yet we have done exceptionally well in weathering these dangers. We live much longer and healthier lives. But we are now pursued by some insidious new foes—from rising temperatures and rising sea levels, to water shortages, to air pollution—which seem to be of our own making. We are threatening the intactness of our home and our own safety.

Until we find an actual Golgafrincham, we need the current one to last forever. Forever is a long time, for people who sometimes only think in terms of election cycles.

The ESG Connection

Absent any ESG influence, we have not been all that good about or consistent in thinking of the planet as our only home and making decisions that help it stay intact, safe, and eternal. In fact, there is ample evidence that suggests we have, for many decades, been rather short-sighted. To quote a comedian, if the planet were a car, we'd drive it like it's stolen.

Fortunately, some far-sighted governments have taken up the cause to provoke more attention to ESG in decision-making. Denmark has declared an end to exploiting its oil resources in the North Sea. The EU has stated its intent to be carbon neutral as a trading bloc by 2050. China has announced its goal to be carbon neutral as a country by 2060.[16]

In addition, the next generation of talent really appreciates ESG. The millennial generation, born between 1980 and 1994, is surrounded by technology and experienced 9/11, the 2008 recession, slow starts to their careers, and high housing and education costs. They stand at the threshold of their peak saving years and maintain a strong belief that the companies in which they invest should go beyond money-making to become part of the solution to environmental and social issues.

Millennials are far more likely to want to work for companies that make strong ESG commitments. Surveys in 2019 by the G&A Institute show that 40 percent of millennials would take a pay cut to work for a responsible employer, and 40 percent had already selected their employer on this basis, compared to just 17 percent of baby boomers.[17]

Capital Markets and ESG

Capital markets have taken note of this tendency and are now exerting very real pressure on the oil and gas industry to declare its goals and intentions regarding ESG. Without that clarity, construction projects can't get insurance, and production can't get funding. Even mighty Alberta, the biggest Canadian investor in clean technology, the undisputed export engine of the country, and whose oil industry is among the most highly regulated globally, has been brought to heel by Wall Street money.[18]

One way to understand how capital markets view energy companies is through market indices. While not perfect (indices reflect more sentiment than just ESG measures), the indices provide an important gauge about market sensitivity. For example, the S&P TSX one-year oil index in 2020 was down 26 percent, while the S&P TSX clean tech and renewable index was up 80 percent.[19]

The problem with oil and gas is that the resource is, by definition, not sustainable. Once extracted, the resource does not naturally or quickly replace itself, like a forest or a fishery. Upstream oil and gas is best viewed as a self-destructive business model. Oil and gas generally is a scale business, and it carries a significant physical footprint. It struggles with its environmental track record in light of its spills and emissions. Some of its legacy designs, such as gas wells that intentionally use methane under pressure to actuate the well mechanics, are fundamentally opposite to ESG ideals.

And in case you think capital shifts are only a concern for energy companies, think again. Tesla—the muscular offspring of an ESG mother and a digital father—saw its stock rise 800 percent in 2020,[20] whereas GM and Ford shares rose by 50 percent, BMW and Volkswagen shares were flat, and Daimler was up by just 20 percent.[21]

International ESG Efforts

Many nations and trading blocs are launching various ESG initiatives, with Europe leading the way. The European Green Deal, announced to much fanfare in December 2019, serves as a sweeping agenda to green up the continent—meaning achieving **carbon neutrality**—by 2050.[22] The deal has broad support—93 percent of Europeans see climate change as a serious concern.

The deal is illustrative of the kinds of changes that will become the norm across many societies in the years ahead. The EU is a large and influential trading bloc, and its regulatory environment is often used as a default standard by companies operating globally.

The Green Deal sets out five transformative goals:

1   to sign into law a binding requirement for member nations to act on climate change and spur investment;

2   to decarbonize the energy sector;

3   to renovate buildings to reduce their energy use and costs;

4   to help European businesses become world leaders in the green economy; and

5   to implement cleaner, cheaper, and healthier forms of private and public transportation.

While the deal highlights key areas in which policies will be developed to achieve carbon neutrality, fossil fuels are so essential to Western life that any new policies will have spillover impacts on the sector. EU member nations will be expected to tabulate the carbon emissions in the entire supply chain, not just those that occur on European soil, lest its economy craftily shift its carbon footprint off the continent, which would defeat the purpose of the policies.

European oil champions such as Shell and Repsol have already announced their goals to transform their own carbon footprints.[23]

THE PRODUCTION and consumption of energy products account for fully 75 percent of the EU's carbon emissions, so the energy industry is going to get the lion's share of the pressure to change. The engineering industry anticipates a dramatic upswing in demand for new energy infrastructure. "Energy transition has been described as the biggest reallocation of capital in industrial history," says Dr. John Pillay, who is senior vice president of digital transformation at Worley.

The energy industry is expected to

- achieve 50 percent reduction in greenhouse gas (**GHG**) from a baseline in 1990 by 2030, and carbon neutrality by 2050;

- connect grids to better utilize renewable energy sources;

- boost energy efficiency;

- decarbonize the gas sector; and

- develop the offshore wind sector.

The Law

The details of the law are not finalized, but the intent is clear. EU member nations will become legally bound to achieve the 2050 target. What countries will do is bring into force the "rules" to achieve the outcomes that they need to be legally compliant. In practice, for example, to obtain a permit, a new business venture will need to demonstrate that its activities will not negatively impact the national target. National industrial policy will need to incorporate carbon neutrality as a key feature. State owned companies, national funding agencies, and sovereign wealth funds will quickly invest accordingly in order to demonstrate progress.

What follows are some of my predictions as to how this will play out.

PLANNED VISIBILITY

Businesses will state in their circulars how they intend to become compliant with the rules. Importers will come under scrutiny to demonstrate that they are not simply exporting their carbon business model overseas beyond the reach of EU law.

By implication, companies that sell into an EU supply chain (energy producers, chemical companies, car parts, electronics, foodstuffs) are going to be pressed into accounting for their carbon emissions and into reducing their carbon footprints.

NO NEW FOSSIL FUEL INVESTMENT

New **greenfield** long-life fossil fuel assets will not be built in Europe. Three decades seems like a long time, but not in the realm of oil and gas, where horizons are often measured in decades. For example, building a new liquified natural gas (LNG) export plant typically takes five to seven years, and such facilities are expected to be operational for at least twenty years, if not longer. Few boards will sanction large shareholder outlays against this backdrop.

Environmentalists now know that a few years' determined delays will be all that's needed to kill off new projects.[24] Investments in new coal, oil, and gas reserves will need to formally address the potential for those assets to become stranded. Any existing reserves on the books that cannot be brought to market promptly may need to be written down.

DIVERSIFICATION OF ENERGY SUPPLY

To stay in business, fossil fuel energy companies will need to become carbon competitive with their zero carbon peers (the sun and the wind), which means achieving net zero emissions. Existing fossil fuel energy businesses (coal power, gas power plants) will need to minimize their carbon impacts, wind up operations, convert to some other purpose such as renewable power generation, and invest in relevant offsets such as carbon capture and storage or plant trees as carbon sinks. Companies will need strategies to progressively transition, or they will face extinction.

The European Green Deal, and many others like it, does not spell the end of the oil and gas industry. There's still a need for plastics, and some transportation fuels (jet fuel) have no alternatives yet. But the pressure to change is suddenly, and legally, very real.

INDUSTRIES BECOME SUSTAINABLE

The European Green Deal sets out a specific goal to develop a true circular economy, whose intent is to address the stress the industry overall imposes on water resources, the generally high emissions footprint of the industry, and the shortcomings of the recycling industry.[25] Products that are harmful and do not allow for reuse, repair, or recycling will eventually be kept out of the market. Companies will need to offer proof of any green claims.

Key industries that will be targeted include steel, cement, textiles, construction, electronics, and plastics. Of these, the oil and gas industry is a big buyer of steel and cement, a consumer of construction services, and the source of virtually all virgin plastic. Oil and gas is also a major producer and consumer of water (steam generation, fracking, reservoir stimulation, drilling fluids). The deal will force reusable and recyclable packaging by 2030, putting downward pressure on virgin plastic demand. New business models based on the sharing economy (goods and services for rent) will be expected to play a role in all industries, including oil and gas.

Some industries will be uniquely impacted. The cement industry has a dual carbon problem—it consumes fossil fuel for heat to create cement, and the cement manufacturing process releases additional GHGs. Replacements for cement products will emerge. Oil and gas will need to stay in lockstep with the cement industry so that new low-carbon cement products also meet the facilities' standards of oil and gas.

The drive to create a circular economy for plastics will impact the demand for both virgin plastics and plastics better suited to recycling. Industry will be compelled to find better methods of recycling plastic. The demand for plastic raw materials is now much more uncertain, as it is dependent on these unknowns.

The construction industry is one of the more laggardly in adopting change, but the Green Deal has now created the conditions needed for the European construction sector to take a lead in transforming for a green future.[26] Others will be left behind.

NEW TRANSPORTATION SOLUTIONS

The deal aims to achieve a 90 percent reduction in GHG from transportation by 2050.[27] Road transport accounts for 72 percent of GHGs from transportation, with aviation and shipping about 27 percent, and rail about 1 percent. Big changes are coming to road transportation, specifically personal transportation.

Interestingly, the deal does not specifically encourage Europeans to drive less, but a few of the proposals look to achieve that outcome.

Some of the Green Deal's suggestions to reduce emissions include the following tactics:

- Impose much tougher vehicle pollution standards. Higher standards motivate the auto industry to shift to electric drive trains (a shift that is already underway) and remove the option for consumers to opt out of electric vehicles (EV).

- End subsidies for transport fuels. Subsidized fuels encourage trucking instead of rail and water trade. Presumably a few markets are still subsidizing fuel for their citizens, though you wouldn't know it from the fuel retail prices in the major European cities.

- Change road pricing. Low tolls encourage road use. By ramping up tolls, the EU hopes to push trade to use rail and water routes. Tolls also make some personal road trips more costly than rail travel, and so encourage more use of public transit.

- Overhaul the vehicle refueling landscape. It is anticipated that an additional million public charging stations will be needed to enable dramatically more EVs. Today's fuel business will need to react, and quickly, likely by adding charging stations and refueling infrastructure for new cleaner fuels (such as hydrogen).

ASSUMING 50 percent of the refining capacity in Europe produces transportation fuels, which is admittedly a pretty blunt
average, the implications on oil refinery infrastructure are severe—
half of the oil refineries in Europe will eventually be surplus to
need. The problem is that this applies to *half of each refinery*,
since a barrel of crude oil typically produces separate portions
of diesel for trucking, gasoline for cars, and kerosene for jet fuel.
It will take a lot of investment in new refinery kit to reformulate
these unwanted products into feedstock for plastics and lubri-
cants, if there is a market.

European oil and gas companies are already sorting out how they
will respond to their new legal requirements. Companies hoping to
continue to sell into Europe will also have to respond, or be prepared
to sacrifice market access.

Getting ESG Right

It's hard to get ESG right, particularly in the energy sector, as the elements are devilishly intertwined. I was schooled on this fact several
years ago when I was invited to Vancouver to meet with representa-
tives of five local Indigenous nations.

There was a proposal at the time to build an expanded oil export
terminal, a subject on which I had some special expertise, in Vancouver
Harbour, their traditional territory.

At the meeting, each leader took the podium for a few minutes to
make their opening remarks. Unfailingly, they called out their rever-
ence for Mother Earth, their reliance on natural fisheries for food, and
their role as stewards of the land. They decried pollution, the loss of
habitat for hunting, and the threat to salmon from an oil spill.

One leader stated, "Under no circumstances will a pipeline ever
be built in our sacred waters," meaning the Vancouver Harbour, one
of Canada's largest and most diversified ports and a gateway to Asia. I
leaned over to the lawyer beside me and whispered, "Why are we here?"

At the break, I learned why. Each leader quietly pulled us away from the others and asked us a set of revealing questions:

- Can young people from my nation find jobs on this pipeline?
- Are there jobs in construction and operation?
- Do pipelines make money? How much?
- What would it take for us to be an owner in this project?
- How do we negotiate a stake in the project?
- What is the approval process for new pipelines?

They talked about the crisis of unemployment in their traditional territories, the anguish of substance abuse, their desire to break free of the chains of poverty. They completely recognized their reliance on petroleum for their snow machines, chainsaws, fishing boats, and heating. The prosperity, comfort, and relief from poverty that oil and gas jobs can provide was acknowledged. Unfortunately, these benefits are perceived to come at the cost of environmental sustainability.

This was the set of circumstances in the lead-up to 2020. European Green New Deal, ESG concerns, capital flight, and oil price collapse. It was a tense time, and although there are arguments that demand will return, and that these deals are far from perfect (or even possible), the future appears to be up in the air.

It seemed things could not get much more challenging.

The Pandemic

It's now a meme to point out that COVID-19 has had a more transformative impact on industry than CEOs, boards, or CFOs. Although anecdotally accurate, the pandemic has really been more of an accelerant on an already-burning fire. Many of the changes put in place in reaction to COVID are now permanent. As John Pillay describes, "The pandemic has accelerated everything, in terms of virtualizing the proposition, globalizing the workforce. It won't roll back." The uncertainties of the virus and vaccination efficacy mean that more changes will come.

COVID as Accelerant

A strange new virus appeared in Wuhan, China, transmitted by the unaware to the unprepared through microscopic airborne aerosols and causing pneumonia-like illness. The failure to contain its spread led to a pandemic that consumed the world for all of 2020 and 2021. The world is visited by weird new illnesses regularly, every decade or so. Recent examples in my lifetime include SARS, bird flu, swine flu, and MERS (also known as camel flu). These are on top of the seasonal flu, which befalls hundreds of thousands annually, with a high mortality rate among the elderly.

COVID-19 has brought home the impacts of virulent diseases to the masses. It is uniquely infectious, possessing a spike protein chain that allows it to hop from cell to cell, person to person, with incredible ease.[28] Where other viruses proved more containable, COVID-19 swept the world in a flash.

Prior to the pandemic, most of us spent precious little time thinking about how viruses impacted our human lives, and just as little time considering the role of viruses and other health threats in our industrial world. Most outbreaks of other illnesses were largely under control, contained by vaccination programs or otherwise. COVID-19 has prompted considerable introspection among boards and management teams about the health and security of their human talent.

At the same time, industry is adding billions of new sensors and digital **automation** into its infrastructure, creating a fertile new landscape for the transmission of computer malware, viruses, and other malicious elements.

The approaches to human health during a pandemic and the management of exposure to computer viruses are curiously similar. Companies are now much more practiced at dealing with computer viruses and are well advised to apply the lessons from their computer exposures to help with surviving this pandemic and preparing for the ones that follow:

- Computer viruses mutate and evolve over time, much like COVID. The idea that the industry can revert to its previous state once COVID is vanquished through immunization is false, as it is with computer malignancies. Industry has invested in permanent changes to cope with computer viruses (training programs, audits, surveillance, armor) and disease prevention will now be added to our permanent defenses as well.

- Older technologies are particularly vulnerable to computer attack, similar to how COVID and other human diseases prey on the elderly and the immunocompromised. Industry already invests special attention on vulnerable older systems as a priority, particularly if those systems are mission critical, as they are with older SCADA systems, sensors, and networks. Industry now must reflect carefully on its insistence that its older workforce return to the office.

- Specialist expertise is required to provide adequate protective services across the range of computer systems in use by industry. The notion that two or three part-time security staff are all that is needed to defend against the **cyber** world is folly. Once hacked, companies typically need a range of services immediately, from brand damage control to technical skills and recovery expertise, and money is usually no object.

- Speed of response is critical to dealing with viruses of the computer variety. They spread quickly, with vigor, and largely undetected. Locking down, tracing the attacks, and quarantining the infected are the same actions that work in dealing with human disease. Having a plan that is frequently tested is key.

One important feature of the human virus world that does not meaningfully exist in the digital world is the information clearing houses that monitor and then share information about human afflictions. There are no equivalents to the World Health Organization and the Centers for Disease Control and Prevention in the computer world. One reason is that computer viruses are often inventions of state actors, and few governments wish to secretly fund a computer attack arm while publicly funding a virus surveillance unit. Another is that successful computer hacks can be very costly in terms of brand damage, stock price impacts, and out-of-pocket recovery expenses. Few companies are willing to go public with their cyber woes.

Companies will be highly dependent on specialist services to help maintain their defenses, and on building cybersecurity into their digital designs as a priority.

The Preparedness Imperative

COVID-19 has schooled the world on the need for emergency preparedness to deal with virus outbreaks. Most nations, with the exception of Australia, New Zealand, and other island countries, did not act decisively at the onset of the virus. This wasn't split along political lines either; both Canada and the US, with liberal Trudeau and conservative Trump, respectively, waited until the eleventh hour to take any action. The disease appeared as far back as December of 2019, according to Canadian military intelligence, and arrived in North America en masse by March of 2020.[29] Still, international border closures and quarantines were not established until April in many countries. Eventually, though, lockdowns were imposed, and the world was trapped at home.

Some nations quarantined entire cities and built hospitals in a week. International borders closed tightly, cruise ships halted tours, and sporting events were canceled or deferred.

Pandemic Tactics

Companies swiftly implemented a set of social distancing measures, such as a work-from-home strategy. This served well those employees principally in commercial roles, including finance, procurement, trading, HR, legal, administration, IT, and other similar service jobs. Entire buildings were closed or were operating in a split shift (half of the employees working from home, the other half more widely spreading out in the office setting).

But not everyone in oil and gas can elect to self-isolate. Field assets need continuous human supervision. Broken equipment cannot just repair itself. Control rooms are confining spaces, forcing operators to work in close quarters. Offshore platforms, where space is at a premium and off-shift accommodations are shared hot bunks, cannot easily meet social distancing targets.

New build, or greenfield, assets under construction are also a challenge. Oil and gas is often located in remote settings and serviced via work camps. (For example, the Fort McMurray area has some 32,000 workers flying in and out.[30] And Western Australia's mining industry is even more dependent on a traveling workforce.) Asset construction planning likely hasn't taken into account the CDC's recommendation that everyone maintain a six-foot (or two-meter) gap between each other.

These blunt-force measures are necessary not only because of how businesses have been designed, but also because of how our health care system works, how our institutional mechanisms have evolved to implement urgent change, and how we view privacy.

Hand-to-Hand Virus Combat

In the heat of the moment, the normal human response to a novel threat is to apply the training one has learned for familiar threats and to rely on proven tools and tactics readily at hand.

ONE REASON the Roman Empire fell was due to a new threat, the lightly armored mounted archer, who was superior in combat against Rome's slow, heavily armed foot soldiers. The Romans unwisely clung to their training and success formula but eventually fell after three hundred years. The early Middle Ages in England were punctuated by lightly armored warriors with shields and spears, fighting on foot. Norman knights on horseback brought the Angles, the Saxons, and the Vikings to heel. In the Second World War, the German tactic of lightning war, or blitzkrieg, swept through country after country for two years, until it met its match in the battle for Stalingrad. The desperation of the Russian people compelled them to try something new—street by street fighting, with constantly moving small bands of fighters able to outmaneuver the mechanized German tank forces and attack the supply line bringing fuel to the front.

This phenomenon, the initial application of yesterday's solutions and tools to unfamiliar problems to see if they work, is what many Western societies and businesses applied to the pandemic. It was likely far more costly than it needed to be.

Over time, the oil and gas sector put into place a comprehensive program of pandemic response, but the solutions adopted are, in the short term, profoundly physical and manual in nature and cost additive.

First, they minimized the number of sites to manage by shutting down some facilities entirely, mandating work from home for suitable roles, and by furloughing employees to reduce the number of workers. Second, they upgraded ventilation systems where possible, bringing them to a much higher standard. Third, they altered site access by adjusting work schedules (which reduces crowding at start- and stop-work times, lunch period, and other breaks), instituting spaced queues at all entry and exit points, restricting access to untested workers and visitors, and mandating quarantines when infections appear. Fourth, they changed how some sites were used by adding more wash and sanitization stations; stepping up cleaning protocols; offering masks, gloves, and face shields; adding vapor barriers in tight quarters such as control rooms; and spacing out employees in offices through split shifts or rented space. Finally, they adjusted movements on sites by better managing shift changes, adding more on-site transit services, and restricting work gatherings.

None of these tactics fundamentally challenged the underlying business models and structures of the oil and gas business. Without more profound business change, these protocols will remain in place, and the degradation of business performance will become permanent.

These added costs now call for solutions that offer long-term relief, and as we'll see in the case studies, digital solutions offer the answer. For example, Senior Vice President of Operations and Technology Jay Billesberger, from NorthRiver Midstream, described how his company needed to do "a huge turnaround [. . .] during the pandemic but couldn't get people working closely together. So, we rolled out a digital permitting system in just three months, from conception to rollout. Permitting went from four hours to half an hour."

Digital Is the Future

It may feel as though the case isn't there for oil and gas to adopt digital. Capital is very hard, almost impossible, to obtain. The demand is sliding ever downward, at least according to the industry pundits. Policymakers are keen to slash and destroy demand as well, banning new petroleum vehicles earlier and earlier.[31] The energy transition is fully underway, with copious spending on new infrastructure programs.[32] Why, given all these headwinds, should oil and gas bother to invest?

The reality of the situation is far less dire, though, when you drill a little deeper. There is a market for hydrocarbons yet. The pandemic and energy transition are two new vectors for demand. In the short term, the post-pandemic era will lock pent-up demand for travel and services, while the energy transition will drive demand for fuel to build new energy sources. Digital promises to capture long-term benefits from short-term demand and will help avoid huge losses from demand destruction.

Rising Demand

As vaccines roll out and people return to work, there will likely be a renewed demand for travel as borders reopen.[33] Although this may take a few years, there will be an increase in traveling internationally and vacationing by the average consumer when it becomes possible to do so. The forecasts vary, but leisure travel will likely return to normal in the years following 2022, given the pent-up demand for tourism, as governments lift restrictions. Time will tell how large the surge will be, but it will come.

Although demand for gasoline may decrease in the long term, jet fuel demand will certainly increase in the next five years. A renewed surge in jet fuel consumption will likely create market opportunities for oil and gas to capitalize, especially in the downstream. This will not replace demand for road-borne vehicles for sure, but air travel will still offer a market for oil and gas for the time being.

Energy Transition Isn't Carbon-Free

The infrastructure projects being launched by governments globally, as well as the energy transition itself, present an opportunity for capturing hydrocarbon demand.

Green energy does not burn hydrocarbons to make electricity, that much is true. But it isn't as if steel, aluminum, concrete, semiconductors, copper wiring, or lithium-ion batteries come from thin air. They all require hydrocarbons, often in the form of kerosene or propane. Not to mention the volume of hydrocarbons necessary for the mining industry to just find all these raw materials. Oil and gas, ironically enough, is very much necessary to build its own replacements.

The amount of energy and capital involved in this project is staggering. One estimate has $15 trillion for the cost of the transition alone; this amounts to having 56 percent of energy demand being met by renewables.[34] Another estimate, by BloombergNEF, suggests up to $130 trillion will be spent globally to achieve 2050 targets.[35] Total replacement, alongside matching the increase in energy consumption, could balloon this number even higher. This results in a large amount of investment in capital projects and infrastructure, all of which will need hydrocarbons.

In short, the transition will entail a temporary increase in oil demand while it is underway, petering out by 2050. While gasoline demand may be gone by then, oil and gas will find business partners elsewhere in the supply chain.

Rise in travel demand after the pandemic and infrastructure and commodity market expansion are just two examples of how the demand for hydrocarbons may not vanish but actually go up in the short term. The fuel demand will return eventually, matched by a spike in industrial fuel demand to build all these infrastructure projects. Oil and gas companies need to move quickly to capture this windfall.

While there are some positives for the industry in the short term, it doesn't mean "business as usual" for oil and gas. Demand *will* disappear eventually. The grid will become much less dependent on hydrocarbons as fuel. Changes will still need to be made to ensure that short-term gains translate to long-term success. This is where digital comes in.

Think Digital

In 2019, digital was viewed in oil and gas as "a future," but not "the future." Other possible futures, notably the status quo, still held sway. Teleworking and teleconferencing on a grand scale were politically impossible in December 2019 but are now part of the fabric for many **B2B** service providers that aim to prosper in the years ahead. As workers and bosses became used to teleworking, they demanded deeper digitalization of the work world. It became clear to many in the industry that digital innovations are one of the very few tools available, including for both energy producers and consumers, to lower costs and improve productivity, reconfigure business to improve its resilience, and meet ESG objectives. Expect to see more budget being applied to strengthen, expand, and evolve the digital foundations of business.

I'M REMINDED of a quote from Milton Friedman.

"Only a crisis—actual or perceived—produces real change. When that crisis occurs, the actions that are taken depend on the ideas that are lying around."[36]

Friedman goes on to say that only in a crisis can the politically impossible become the politically inevitable.

I should point out that while some enterprises (Airbnb) will be asset-light, almost purely on a digital platform, not all businesses will have that option, and for now, oil and gas doesn't. We still need fuel to grow food, energy to provide heat and light, power to manufacture clothing, and petroleum for transportation. Physical assets continue their central role in energy systems. Digital's role is to help, not replace.

Here are some other digital insights that have gained life because of capital pressures, demand changes, ESG agendas, and the pandemic.

DATA TRULY IS THE NEW OIL

Centralized workers in an office and shoulder-to-shoulder operators in a control room can, in their various ways, cope with poor quality data about assets. But remote workers, at-home bosses, and robotic machine-based businesses incur heavy prices for poor data quality. The pandemic even prohibits impromptu site visits to survey installed equipment as a crutch for faulty records. High-quality data is now in high demand. Interest in data as the "product" rises.

NETWORKS ARE CRITICAL

Many at-home-workers are discovering that their home networks cannot cope with Zoom and Netflix, kids and adults, all competing for the same limited **Wi-Fi** resource. Companies have long resisted getting involved with network infrastructure outside the firewall, but now the productivity of the enterprise is dependent on how robust the home is, as well as the networks to the remote edges of the business. Expect to see telecom companies stepping up to deliver beefier networks.

SENSORS WILL REPLACE HUMAN SENSES

Instead of having workers traveling to sites to check equipment, operators can deploy cameras equipped with **visual analytics** to keep eyes on assets. Visual analytics is part of the digital field that includes facial recognition technology but is rapidly being applied to a huge range of industrial uses. One large construction company uses its overnight security cameras to identify the arrival of parts, equipment, and cement to its construction sites and to alert crews. Used in this way, cameras reduce carbon, lower costs, and keep employees out of hospital. Camera sensors are but one example—other sensors measure sounds, smells, vibrations, temperatures, pressures . . . the list is endless.

EDGE DEVICES WILL PROLIFERATE

The power of distributed inexpensive sensors coupled with **cloud**-enabled machine interpretation of sensor data will unlock demand for **edge device**s that continuously self-monitor and do not require constant human supervision. The first uses will be to run remote operating equipment in the field, with the occasional check-in as a satellite network link comes around.[37] As the industry grows more comfortable with the reliability and trustworthiness of these devices, they will take on more of the routine field supervision role. **Drone**s, the most rapidly developing class of edge devices, are advancing quickly because of low-power chips and better battery technologies.

MACHINE-CENTRIC BUSINESS MODELS WILL PREVAIL

Business models that are dependent on people working in forced close contact are in peril. Workers are justifiably alarmed at the prospect of working conditions dependent on confinement and proximity. Oil and gas has invested in many rigid technologies that will remain human-centric for operations and maintenance, but edge devices that can run safely and reliably, and are virus-proof, are the future. The rise of the machine-based business model is here. Expect to see a big jump in interest in algorithms, **machine learning**, artificial intelligence (**AI**), and **autonomy** that contribute to keeping equipment working longer and harder without human supervision.

BROWNFIELD ASSETS WILL GET NEW LIFE

The majority of oil and gas infrastructure, from wells to gas stations, predates ESG concerns and the digital era. These assets now serve as a drag on the ability of companies to achieve much progress on their ESG commitments because they are so resistant to change. On the other hand, they can be data-rich assets because of their connections to SCADA and other monitoring systems, and they gain hugely from the analytic possibilities of machine learning and AI. Such tools can help improve the quality of legacy data so that it yields better analytic outcomes, as well as helping in conducting better analytics on new data. Better analytics leads to better operations decisions that include ESG targets. In time, **brownfield** assets can be managed more tightly and in conformance with ESG goals.

CARBON MEASUREMENTS WILL IMPROVE

Brownfield and greenfield assets will be material carbon sources for the foreseeable future, which means the industry will need to carefully track its carbon position so that it can make appropriate positive offsets. Today, carbon measurements tend to be from engineering principles, whereby a given asset, designed to run at a certain level with a fuel of known characteristics, has an estimated carbon output.

However, assets leak, valves drift out of calibration, and different gases have wildly different impacts. Because of scale effects, minor variances in carbon measurement accuracy can add up to huge absolute differences from engineering estimates.

Digital tools such as edge sensors and satellite imagery interpretation can help provide near-real-time, continuous monitoring of actual asset carbon impacts by detecting vapors and recording measurement data with low latency in easy-to-access cloud databases.

SUPPLY CHAINS WILL BE RECONFIGURED

With in-person collaboration now a risky undertaking, and the proven ability of conferencing tools like Zoom and Teams to compensate for the loss of the in-person experience, expect to see collaboration tools—joint document editing (Google Docs), shared work tasks (Trello, Teams), team communications (**Slack**)—to be deployed more enthusiastically in the supply chain, with both contractors and suppliers. Global networks allow service companies to offer real-time asset supervision from anywhere in the world.

SUPPLY CHAINS WILL BECOME TRANSPARENT

The supply chain for oil and gas is long and complex. Tracing products throughout the supply chain to provide the assurance that the products were sourced from ethical suppliers with meaningful ESG practices is fast becoming a requirement of global brands. This is already very pronounced in consumer goods, pharmaceuticals, and many food products, and has now come to chemicals.

Digital innovations provide better tracking and tracing of fluids, gases, and commodities throughout the supply chain, given the chain's high level of fragmentation, multiple handoffs, discrete services, frequent changes in control, and high regulatory burden. Tools like **blockchain** are now very handy in helping deliver the transparency that supply chain participants need to assert to their ESG metrics.

CAPITAL ACCESS WILL ADAPT

Traditional capital markets are now skittish about lending to the fossil fuel industry, and regulators are forcing lenders to be transparent about their exposures to energy market transitions. At the same time, digital innovations are also creating new pools of capital that, until regulation catches up, may be used to finance the industry, by fractionalizing asset ownership, **tokenizing** oil production, and settling trade. It's a good bet that someone somewhere has already concluded an oil transaction using bitcoin. Expect the industry to progressively explore these new financing tools to assist with its operations.

IT'S BECOME clear that the adoption of new ways of working enabled by digital tools is a critical pathway to lower costs, boost productivity, and unlock new business models. According to Jay Billesberger, "digital takes noise out of your system. Everybody immediately knows what the baseline numbers are. You manage by exception."

Moreover, it turns out that digital innovations are *the* key solution to solving the problems of the pandemic *and* the cost challenges of the industry. And they've been in front of us for a couple of years now.

Those that embrace digital innovations can expect cost reductions of 20 percent or more and productivity gains of 20 percent or more. Some companies, like Repsol, have staked their future on achieving carbon neutrality, which can only be done if the work processes that generate carbon are overhauled, in line with solving pandemic challenges, and costs are lowered. Digital is the way forward.

KEY TAKEAWAYS

Here are a few key takeaways from the forces of change facing the industry:

1. The transition to a more diversified and rebalanced energy mix is now in motion and measurable.

2. Future demand for fossil fuels for transportation is very much uncertain, but the demand for plastics and petrochemicals is still robust.

3. Environment, social, and governance factors now weigh very heavily in making decisions about energy sourcing and consumption.

4. The climate legislation from the European Union is the global pacesetter for the future of fossil fuels.

5. Capital markets now decisively favor digital companies.

6. The pandemic taught the energy industry that it can change quickly when necessary.

7. Digital tools that were in place but underutilized demonstrated their value during the pandemic.

8. The oil and gas industry is still far from capturing full value from digital innovation.

Notes

1   BP, "Energy Outlook: 2020 Edition," 2020, bp.com/content/dam/bp/business-sites/en/global/corporate/pdfs/energy-economics/energy-outlook/bp-energy-outlook-2020.pdf.

2   "Responsible Investment Annual Briefing Updates," Shell, April 16, 2020, shell.com/media/news-and-media-releases/2020/responsible-investment-annual-briefing-updates.html.

3   BP, "Energy Outlook 2020."

4   BP, "Statistical Review of World Energy: 2020," 2020, bp.com/content/dam/bp/business-sites/en/global/corporate/pdfs/energy-economics/statistical-review/bp-stats-review-2020-full-report.pdf.

5   "Crude Oil Prices—70 Year Historical Chart," Macrotrends, accessed May 5, 2020, macrotrends.net/1369/crude-oil-price-history-chart.

6   Organization of the Petroleum Exporting Countries, "OPEC Annual Statistical Bulletin," Scott Laury, ed., OPEC.org, 2020, asb.opec.org.

7   BP, "Statistical Review of World Energy: 2020."

8   Olga Yagova, "Saudi Arabia Floods Markets with $25 Oil as Russia Fight Escalates," Reuters, March 13, 2020, reuters.com/article/us-oil-opec-saudi-idUSKBN21O22H.

9   US Energy Information Administration, "Early 2020 Drop in Crude Oil Prices Led to Write-Downs of U.S. Oil Producers' Assets," EIA, July 27, 2020, eia.gov/todayinenergy/detail.php?id=44516.

10 Dr. Werner Vogels, "The Story of Apollo—Amazon's Deployment Engine," All Things Distributed, November 12, 2014, allthingsdistributed.com/2014/11/apollo-amazon-deployment-engine.html.

11 Alex Kimani, "BlackRock Is Turning Up the Heat on Oil Companies," OilPrice.com, January 27, 2021, oilprice.com/Energy/Energy-General/BlackRock-Is-Turning-Up-The-Heat-On-Oil-Companies.html.

12 Jeff Loucks, James Macaulay, Andy Noronha, Michael Wade, and John T. Chambers, *Digital Vortex: How Today's Market Leaders Can Beat Disruptive Competitors at Their Own Game* (Lausanne, Switzerland: IMD—International Institute for Management Development, 2016).

13 Geoffrey Cann and Rachael Goydan, *Bits, Bytes, and Barrels: The Digital Transformation of Oil and Gas* (MADCann Press, 2019).

14 International Energy Agency, "Digitalisation and Energy: Technology Report," OECD/IEA, November 2017, iea.org/reports/digitalisation-and-energy.

15 A planet that is "doomed" and sends its population in ark-like ships to colonize new worlds. Unfortunately, the only survivors are management consultants and telephone sanitizers. From Douglas Adams, *The Restaurant at the End of the Universe* (London: Pan Books, 1980).

16 "Denmark Set to End All New Oil and Gas Exploration," BBC News, December
4, 2020, bbc.com/news/business-55184580; "2050 Long-Term Strategy,"
Climate Action, European Commission, November 23, 2016, ec.europa.eu/clima/
policies/strategies/2050_en; Matt McGrath, "Climate Change: China Aims for
'Carbon Neutrality by 2060,'" BBC News, September 22, 2020, bbc.com/news/
science-environment-54256826.

17 G&A Sustainability Highlights, "Millennials Really Do Want To Work for
Environmentally-Sustainable Companies, According to a New Survey of
Large Company Employees," Governance and Accountability Institute,
February 23, 2019, ga-institute.com/newsletter/press-release/article/
millennials-really-do-want-to-work-for-environmentally-sustainable-companies-
according-to-a-new-su.html.

18 "Oil-Rich Alberta Seeks Ways to Go Green," *The Economist*,
December 3, 2020, economist.com/the-americas/2020/12/05/
oil-rich-alberta-seeks-ways-to-go-green.

19 Index data gathered from Yahoo Finance as of February 2021.

20 Index data gathered from Yahoo Finance as of February 2021.

21 Stock price data gathered from Yahoo Finance as of February 2021.

22 "A European Green Deal," European Commission, ec.europa.eu/info/strategy/
priorities-2019-2024/european-green-deal_en.

23 "Responsible Investment Annual Briefing Updates," Shell, April 16, 2020, shell.
com/media/news-and-media-releases/2020/responsible-investment-annual-
briefing-updates.html; Press release, "Repsol Will Be a Net Zero Emissions
Company by 2050," Repsol, December 3, 2019, repsol.com/en/press-room/press-
releases/2019/repsol-will-be-a-net-zero-emissions-company-by-2050.cshtml.

24 "Keystone XL Pipeline: Why Is It so Disputed?" BBC News, January 21, 2021, bbc.
com/news/world-us-canada-30103078.

25 "A European Green Deal."

26 Peter Evans-Greenwood, Robert Hillard, and Peter Williams, "Digitalizing the
Construction Industry: A Case Study in Complex Disruption," Deloitte Insights,
February 26, 2019, deloitte.com/us/en/insights/topics/digital-transformation/
digitizing-the-construction-industry.html.

27 "A European Green Deal."

28 Dr. Barry Robson, "COVID-19 Coronavirus Spike Protein Analysis for Synthetic
Vaccines, a Peptidomimetic Antagonist, and Therapeutic Drugs, and Analysis of
a Proposed Achilles' Heel Conserved Region to Minimize Probability of Escape
Mutations and Drug Resistance," *Computers in Biology and Medicine* 121 (June
2020): 103749, doi.org/10.1016/j.compbiomed.2020.103749.

29 Murray Brewster, "Canadian Military Intelligence Unit Issued Warning about
Wuhan Outbreak Back in January," CBC News, April 10, 2020, cbc.ca/news/
politics/coronavirus-pandemic-covid-canadian-military-intelligence-wuhan-1.55
28381.

30 "Passenger Statistics," Fort McMurray International Airport, accessed March 24, 2021, flyymm.com/passenger-statistics.

31 Jay Ramey, "Canada to Ban New Gas-Engined Car Sales by 2035," Autoweek, June 30, 2021, autoweek.com/news/green-cars/a36888320/canada-to-ban-new-gas-engine-car-sales.

32 "Fact Sheet: President Biden Announces Support for the Bipartisan Infrastructure Framework," The White House, June 24, 2021, whitehouse.gov/briefing-room/statements-releases/2021/06/24/fact-sheet-president-biden-announces-support-for-the-bipartisan-infrastructure-framework.

33 Jeremy Bogaisky, "What's Ahead For Airlines and Aviation In 2021," *Forbes*, December 29, 2020, forbes.com/sites/jeremybogaisky/2021/12/29/whats-ahead-for-airlines-and-aviation-in-2021.

34 Irina Slav, "The True Cost of the Global Energy Transition," OilPrice.com, November 9, 2020, oilprice.com/Energy/Energy-General/The-True-Cost-Of-The-Global-Energy-Transition.html.

35 "Emissions and Coal Have Peaked as Covid-19 Saves 2.5 Years of Emissions, Accelerates Energy Transition," *BloombergNEF*, October 27, 2020, about.bnef.com/blog/emissions-and-coal-have-peaked-as-covid-19-saves-2-5-years-of-emissions-accelerates-energy-transition.

36 Milton Friedman, *Capitalism and Freedom* (Chicago: University of Chicago Press, 2020).

37 Alex MacGregor, "Husky Energy Deploys Ambyint's AI Technology across Rainbow Lake Field and Demonstrates Impact That IIoT Can Have on Operational Excellence in Oil & Gas," Ambyint, November 13, 2018, ambyint.com/resource-item/husky-energy-deploys-ambyints-ai-technology-across-rainbow-lake-field-and-demonstrates-impact-that-iiot-can-have-on-operational-excellence-in-oil-gas.

THE DIGITAL
FRAMEWORK

"Don't call it technology. That sounds bleeding edge. Use less threatening language. Call it a 'mechanical solution.'"

CORY BERGH, VP, NAL Resources

O F THE hundreds of digital technology types vying for market attention, a relatively small number appeal to the oil and gas industry. The industry is averse to cutting-edge technologies, and generally only those with a proven track record of success within the industry can penetrate and capture market share. It is also difficult and challenging to adopt new technologies given the size, scale, and complexity in oil and gas. Unlike the consumer sector, which features dozens of unicorns (startups with billion-dollar valuations), oil and gas has few if any in this vaunted category.

Nevertheless, there are several digital technologies that are becoming increasingly prevalent. These priority digital technologies now appear throughout the industry in a variety of areas and applications, and shortages of talent are already apparent.

But what drives some technologies ahead of others, how do these technologies converge to amplify their impact, and which ones are leading in the race to digitize the industry? This chapter presents the Digital Framework, a simple model that can be sketched out on a napkin. It can be used to explain the must-have digital technologies and skills, how they relate to one another, and the value equation based on the laws of digital.

Digital technologies work together to create value greater than they can create individually. Here is a simple framework for approaching digital innovations, and it applies throughout oil and gas.

The Digital Framework has three layers:

- business capacity
- digital core
- digital foundations

Business Capacity

The top layer is business capacity, or the capacity to behave and make decisions in a manner consistent with the pace of change in digital. Many companies, not just in oil and gas, buy digital tools or solutions and conclude that they are digitized. They *do* digital. But to *be* digital means working differently. Being digital is the hard part.

Being digital means a significant mindset shift and the adoption of new ways to manage change, develop skills, organize talent, and make decisions. I have dedicated an entire chapter, Chapter 4, to addressing questions around change management and talent.

Being digital also means adopting new ways of working that are consistent with and, more importantly, can keep pace with the rate of change in the underlying digital technologies. The first new way of working is called **agile**, whose methods augment and, in many cases, replace traditional sequential work practices. Agile originated in the technology industry as an iterative approach to developing new software or hardware, and it is based on how much work can be done within a time frame, rather than setting a distinct scope and work plan.

The second new way of working is called **UX**, or user experience. Digital innovators concentrate dramatically more energy on making their solutions very easy to use—so easy that training is practically unnecessary. The interface between the user and the technology is intuitive and seductive, and it exploits the kind of online game techniques that make work irresistible, fun, and even mildly addictive.

As the case studies will show, creating the capacity to work in this fashion, given the hard assets of oil and gas, is what sets leaders apart.

DIGITAL FRAMEWORK

BUSINESS CAPACITY

For "being digital," new ways of working

**PEOPLE & CHANGE MANAGEMENT
(CHAPTER 4)**

AGILE & USER EXPERIENCE

DIGITAL CORE

Unique to specific companies, creates
signature ways of working

DATA

Industrial Internet of Things

Artificial Intelligence

Robots

DIGITAL FOUNDATIONS

Applies to all companies and organizations

CLOUD COMPUTING

BLOCKCHAIN

**ENTERPRISE SYSTEMS:
CYBER, PLATFORMS, APPLICATIONS**

Agile Ways of Working

Agile methods—shorthand for new ways of conceptualizing, designing, building, testing, and deploying software—came about because the previous ways of working in software were incompatible with the pace of change of technology. Taking months to document and sign off user requirements for a new system before embarking on an even longer marathon effort to build the system doesn't work when the requirements change or the technology improves.

Software engineers needed new ways to accelerate testing, deploy software changes more quickly into production, streamline the process for solving user issues, and, most importantly, develop software solutions with greater speed and with a better fit with the users' actual needs. Some of their important innovations include

- new roles for human participants, including product managers, **scrum** masters, and customers;

- basing work effort on a set deadline, or time-boxing, rather than on the scope of the project to yield deliverables or outcomes on a frequent basis;

- iterating development to create multiple precisely wrong but directionally correct versions of solutions that eventually zero in on a workable answer;

- no-code or low-code tools for building solutions that eliminate the need for specialized programmers altogether; and

- automation of testing to allow for more comprehensive testing, earlier discovery of errors, and a reduced demand for human testers.

Agile methods have now entered areas outside of software development, including the oil and gas industry. All of the case study companies mentioned use agile ways of working to one degree or another. But it's equally important to note that none had completely abandoned their traditional techniques.

Agile methods work very well when the solution is uncertain or the requirements are hard to pin down. Agile is not well suited to those situations where the outcomes are very clear or well understood. As one interviewee put it, you do not want engineers running trials or producing minimally viable products where the product is a highly pressurized or energized asset, such as a boiler or a motor.

Some agile techniques, such as the daily **stand-up** meeting, are being adopted throughout oil and gas, even around the CEO table, precisely because they are so well suited to operating in environments of uncertainty.

The User Experience

The second important feature of business capability is the user's experience with the solution. A well-designed UX has proved to be singularly key to growing adoption of digital tools. There are six levels of UX design from consumer technology that have a place in our more industrial world.

USE

Digital companies obsess over how their technologies are actually used. The placement of a single button, its size, color, and font generate hours of heated discussion. Similar debates frame the design of menus, screen layouts, features, accessibility, security, interactions, notifications, and many other interfaces. Analysts pore over videos of user activity, finger movements, eye focus, facial expressions, hovering time, and navigation. The goal is speed—helping the user get what they want as quickly as possible.

THE DISTINCTION between an engineering design and a user-oriented design is apparent in the world of TV remote controls. The remote from one of the big TV makers has a button for every feature, creating this oversized, complicated, thick, and heavy controller. The Apple TV remote, on the other hand, is much simpler, with a hidden trackpad and buttons for volume control, power, microphone, stop-start, and menu. That's it. It's been that way for a decade.

The vast majority of technologies on offer to oil and gas lacks this attention to the needs of the user. Most are oriented to the needs of the asset, providing gauges and displays of the asset's current activity readings, such as operating temperature and pressure. The human operator is often an afterthought. This explains in part why new business solutions and technologies developed without much consideration for the ultimate human user struggle to gain much uptake and are greeted with ambivalence and occasionally hostility.

PERSONALIZATION

Online games highlight the second feature of UX, which is the level of personalization that the user can control. The greater the level of personalization, the deeper the attachment the user exhibits towards the solution. Game players can alter their avatars' hairstyles, eye colors, body shapes, genders, and, in some fantasy games, become a mythical creature (such as an elf, troll, or goblin). Appearances are adapted further through costumes, garb, armor, tools, and equipment. Players personalize their challenges in the game, such as completing a specific route within some constraint, such as time or speed. Badges, rewards, tokens, advanced tools, and special skills all accrue with time and skill attainment. Sound effects and music help create the emotional tone and encourage participation.

Personalization settings are very important to the user. Note how when **smart**phone makers issue the latest version of their phones, they are very careful to include the ability to migrate all the personalization settings to the new phone. The disruption of losing a highly personalized experience because of migration is a critical barrier to adoption and upgrades.

The ability to personalize oil and gas systems varies tremendously. Legacy technologies often have no personalization features. SCADA systems are designed to the constraints of the hardware and not with the flexibility of the operator in mind. A tank gauge displays what the engineer designed it to display, in the layout defined by the physical screen shape. Personalization is minimal.

Well-built digital technologies, even for the industrial world, now feature some level of personalization by allowing users to choose their dashboard contents, screen layouts, notification choices, and screen colors.

ENGAGEMENT

The next level of UX with digital technology is the level of engagement. The bar for engagement is set by consumer products, such as Google Search, iPhones and Apple Watches, Android apps, Facebook, and Uber. These solutions are often so easy to use that they don't even offer training in their features and functions. Children can pick up a tablet and quickly make sense of it.

We fully appreciate and value the features of these products that make them mildly addictive, such as their recommendations for additional content. We willingly participate in flagging material that we deem valuable, by pressing the like button (or the sad face emoji), and we encourage their spread by sharing it with our networks. We willingly offer our feedback in the comments fields. Engagement is an important indicator of how deeply the consumer is emotionally invested in the experience.

Many corporate industrial digital technologies have adopted some of these engagement features, starting with a feedback or comment box for a system. Large collections of data, or libraries of shared solutions or code, can capture user likes, which then help with ranking their utility based on actual usage.

CONNECTION

The next level of UX is the ability to meet and connect with other users. Industrial technology companies have always relied on user groups and conferences to help spur upgrades, sell additional modules, attract new customers, and gather feedback for research and development (**R&D**). These gatherings became virtual during the pandemic, although the quality of the online events was very uneven.

The digital equivalent is the online community of users, which is often launched during the startup phase and is, for cost reasons, entirely virtual. Communities are harnessed to help solve a broad range of problems, from answering questions to finding fixes to technical issues. Digital companies work hard to nurture their communities of users through information exchanges, blog articles, virtual gatherings, and live-stream instructional sessions. The pandemic actually boosted the impacts of these communities.

Industrial technology is still very much behind in adopting these kinds of community-building features found in the digital world.

COLLABORATION

The next level of the experience is collaboration. In the online game world, instead of playing against a machine, players form teams with varying skill levels and learn to work collaboratively within the confines of the game. They progress through skill rankings, compete against other players, acquire yet more advanced tools and skills, collect badges of accomplishment, and amass rewards and tokens redeemable for virtual or real goods. Gaming tournaments are a fast-growth extension of this phenomenon.

Energy infrastructure owners and operators are understandably wary of introducing unrestrained competitive behaviors into their workforce. A performance measure to increase the speed of work can sometimes encourage employees to cut back on safety or on community engagement, to the overall detriment of the business. Competing against one's own performance is often the safest and most productive way forward. Collaboration across companies can run into anti-trust rules, and as a consequence collaboration is a more controlled affair.

CREATION

The highest level of UX is creation. For example, YouTube offers YouTube Studio, which allows, among other things, content creators to edit recordings of their livestream videos just after streaming has ended.

The industrial equivalents are the platform solutions that incorporate innovations like software development kits that let the user develop their own **microapps** or native applications that they can use for their own purposes or can offer to others in an App Store construct.

The risk of unleashing these creative tools in the industrial world is that developers find themselves building solutions and not investing in the user experience!

Digital Core

The core capabilities include four key elements:

1 data
2 the industrial Internet of Things
3 AI and machine learning
4 autonomy and robots

Data (1), which is generated by things (2), is interpreted by artificial intelligence (3) and applied by robots (4). Applying this formula creates signature ways of doing business that are distinctive, hard to copy, and enduring.

It all begins with data. Every business has data, produces data, and consumes data. That data is specific to a process, an asset, or a business. And digital is all about data. Data is now utterly central to the oil and gas company of the future and is positioned on the framework at the center.

Next in the framework is the industrial Internet of Things (IIoT)—the devices that generate data and allow remote access and control of assets. The **IoT** is already widespread in the consumer world. Home management devices, smart switches and thermostats, cameras, toasters, even barbecues have internet-enabled probes so you can know just when your smoked turkey achieves peak smokiness. These devices have their inherent risks and conveniences, although people have been charmed by the novelty.

To read, interpret, and manipulate all the data, we require AI and machine learning. AI refers to the capacity for machines to perform the cognitive tasks typically associated with human beings, such as visual interpretation, language translation, or creative expression. More precisely, it is a series of programs and algorithms that interact to perform a specific task.

Finally, robots apply the data to do real work, replacing humans in the office and the supply chain, as well as in costly, dangerous, and repetitive tasks. Autonomy in the digital framework refers to artificial intelligence or machines that have the capacity to execute commands or instructions without direct human authority.

Every company, every team, every unit, creates their signature ways of working by combining and configuring around these four key elements. Speed is of the essence in creating these signature ways of working because of the combined impacts of the two laws of digital.

Moore's Law

Moore's Law was coined by Intel engineer Gordon Moore.[1] He observed that the density of transistors on a circuit board doubled every eighteen months. This growth rate phenomenon, a doubling per unit time, applies to computer chips that are growing at an exponential rate in capability, but falling in size and cost at a similar rate. It also applies to the software and algorithms that run off of computer hardware. There are three kinds of chips that are growing following Moore's Law:

1   data chips where we store data;
2   computer chips that do computations on the data; and
3   communications chips that move the data and the math results around.

These chips are shrinking in size, becoming ever less costly and power hungry, while growing in throughput, speed, and capacity. In business terms, Moore's Law equates to a compound annual growth rate of 40 percent.

Metcalfe's Law

Metcalfe's Law is attributed to Robert Metcalfe, a researcher in economics at the time of the deregulation of industry in the UK during the 1970s.[2] His law states that the value of a network is proportional to the square of the number of connected nodes on the network. Network assets become increasingly valuable and powerful the more nodes that are attached.

COMBINED, THESE twin laws cause digital innovations built on computer hardware and software to evolve very quickly, and they compel individuals and companies to adopt them at accelerating rates.

Digital innovators are looking to build networks that produce and consume data. Mathematically, they're trying to grow N at 40 percent compounded because it's worth N-squared.

Digital Foundations

The third layer in the Digital Framework comprises the digital foundations. The foundations are the digital table stakes. Without them, you're not in the game. Cloud computing stores the data, provides analytics on demand, and hosts the AI engines, machine learning algorithms, and even some robots. The cloud enables new disruptive business models and provides the means to exploit network effects.

Next, blockchain technology gives us trust over the data, the sensors, the algorithms, and the **bots**. In the future we will no longer rely on people and complex processes to provide the trust we need. It will be delivered through blockchain technology.

Finally, the digital foundation incorporates networks that allow connectivity with the services in the Digital Framework, cyber protection over digital assets, and the enterprise resource planning or commercial systems to carry out a broad range of business services.

THAT IS the model—business capacity, digital core, digital foundations. Try sketching it out on a napkin or relating your latest digital project to it. It will prove very helpful in explaining your digital agenda and in helping your organization move beyond doing and buying digital to being digital.

With that model in mind, it's time to dive deeper into the individual elements and where they sit currently in the industry.

Digital Maturity

The elements that make up the Digital Framework are not equally "mature" in terms of their adoption across the industry. Some are very mature—the case study companies all highlight how cloud computing is now solidly part of their business infrastructure—while others, like blockchain, are still nascent. The S-curve model of maturity is helpful in guiding decision-makers, technologists, and regulators on the deployment of these solutions across the industry.

The S-curve is routinely used as a narrative pathway for technology acceptance and maturity. This is based on the **"Technology Life Cycle"** by Nikolai Kondratiev.[3] My model is narrowly focused on how a technology is adopted within industry, particularly oil and gas.

The x-axis is time, and the y-axis is maturity. Maturity refers to the degree, extensiveness, and proliferation of a technology. In this instance, it is focused exclusively on oil and gas. Early in a technology's appearance in a company or industry, there is a long period of suspicion, a lack of use, and a lengthy series of proof-of-concept trials. Once the use cases and value for novel technologies begin to emerge, and if the use cases are sufficiently compelling, the technologies enter into a period of fast rising acceptance and development. This part of the curve also represents where the "hype" for technology reaches its peak. As they become part of the normal fabric of regular operations, these technologies become "mature" and no longer show the same rapid growth rates.

This framework shows which technologies are growing fast, which ones are about to enter this rapid growth phase, and which ones have left it. Depending on your strategy, you may wish to pursue novel technologies just at the cusp of growth, so you can capture the greatest amount of value. If you are concerned about unwelcome disruption to your business, technologies that are in growth mode may best be avoided.

Data

Data, the first component of the core capabilities, is now completely central to the industry. According to Executive Vice President Patrick Elliott, from Jupiter Resources, "if we get the best people, give them the best information, empower them to participate in the decision-making, we would outperform the market because we would

DIGITAL MATURITY

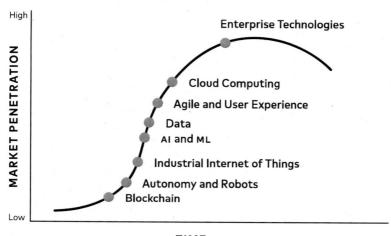

have the full capacity of our people engaged. To do so, they needed full access to data."

In a digital world, data is an asset, a currency, the fuel for work, and, increasingly, pure value. The oil and gas industry has plenty of data to work with, and as will be covered in Chapter 3, on business models, entirely new businesses are starting to emerge based on data. When it comes to managing, controlling, vetting, and applying data, there are still considerable opportunities available, and not just in empowering other digital technologies. For example, as John Pillay puts it, "in a project environment, you tend to create thousands of little islands of data every time you create a project."

Some in the industry have made huge progress in approaching their data very differently. The case studies will highlight several examples, but the majority of organizations still struggle with overcoming their legacy approach to data:

- Most companies lack a clear vision from management about the role and importance of data to the company, regardless of whether it is in upstream, downstream, or in services.

- Industry software product designs tend to imprison data inside specific products and thwart efforts to share. This is viewed as a cost of business and not solvable.

- When purchasing new software, department heads and technical experts tend to overvalue analytic functions and undervalue its data features.

- Companies operate in a landscape of isolated and inaccessible information islands or silos, with multiple definitions of such basic things as gas wells and compressors.

- Excel is the default tool to solve all manner of data-related problems, such as facilitating the exchange of data between two incompatible systems, manipulating data, and visualizing data via charts and graphs.

- Accounting rules continue to treat data as a cost to be managed rather than an asset to exploit.

- Workers lack clear accountability for data, and managers struggle to articulate the capital requirements and benefits from better data management practices.

- Businesses have few performance metrics that are specific to data quality.

- Internal competition for capital between divisions unintentionally thwarts data sharing, collaboration, and mutual support.

And now, a whole new wave of challenges is upon the sector and is commanding attention:

- How will the industry tackle the requirement to identify and then trace its carbon emissions? This is a net new requirement.

- What about the interest on the part of services companies and technology providers to retain copies of data they generate via the sensors they sell or the services they provide? The case studies suggest that considerable value is at stake.

- How, with its reluctance to share data beyond its fence line, will the sector rationalize the opportunity presented by machine learning solutions? The more data these solutions have, the better they are at supporting decisions with quality analysis and recommendations.

- With so much manual manipulation of data, how will the industry be able to convince skeptical capital markets that its data is accurate, complete, and truthful?

The Industrial Internet of Things

The world of IIoT is rapidly growing and, in new energy installations, is becoming the default system choice.

Supervisory control and data acquisition is the legacy industrial design for near-real-time data collection and interpretation from remote assets. SCADA systems are exceptionally reliable (no blue screens of death), rugged, and run around the clock without fail. Without them our current world of energy simply doesn't work.

But SCADA is not well suited for a digital world:

- Integrating with a SCADA system is expensive, which is a problem in a digital world that is all about integration.

- Older versions of SCADA are not enabled for the internet, which limits the ability to manage them remotely.

- Many SCADA systems rely on their obscurity for their security, which is a flawed strategy in a world with robotic hackers.

- SCADA systems are designed for control, not analysis, which is misaligned for a machine learning world.

Legacy SCADA systems are not up the task, but that doesn't spell doom for asset management. Newer sensor technology is more than up to the challenge.

IIoT versus Edge

In the industrial world, IIoT comes in two main forms. The first form are sensors (often bundled into controls units) on a work site that collect data on operating assets and transmit that data to a control center that is also typically on-site. IIoT is different from SCADA in that IIoT devices can both send and receive data or instructions; the sensor can be controlled remotely; some analytics can often be done directly on the sensor.

The second form is called an edge device. Such devices are often located very far from a control center, frequently due to limitations in network connections, or are logistically complex to reach. To be useful, an edge device

- houses some onboard analytics horsepower;

- has connectivity via a network to share its data and calculations;

- can run without a continuous network connection;

- can accept upgrades and software repairs over the network rather than through a human site visit;

- must be secure from cyber activity; and

- must be safe and reliable in hostile environments constrained by power, light, heat, water, and climate.

Edge devices are found throughout oil and gas. In the upstream, the edge might be a drilling rig that is grinding through rock, a storage tank that holds fluids, or a well that is producing hydrocarbons. For midstream companies, the edge is a pump pushing fluids or gases down a pipeline, a generator producing power, or a truck hauling water. For a downstream company, the edge might be an unmanned retail site or an airport in the bush.

THE PERCEPTIVE might note that a modern smartphone, with the equivalent computing horsepower of a 1970s supercomputer, is a fine example of **edge computing**. People do real work, like inspecting an asset, and their phones provide network connectivity, apps for recording observations, storage of the data, continuous upgrades, security from hackers, and tons of other useful features. Some are even waterproof.

The Edge of Upstream

Of all of the segments in oil and gas, the upstream is the ideal candidate for edge computing. Workers agree, according to Jay Billesberger: "Why do I have to climb up and calibrate this old clunky meter every month when the digital ones will not only send me the data, but they don't require recalibration?" There are many reasons why edge will work so well for the upstream:

1 Much activity in the upstream takes place beyond the reach of modern telecom networks, which limits the ability to use cloud computing.

2 Remote sites are frequently already wired up with SCADA systems. The business case to supervise and control the edge with computers is already sound.

3 Modern analytic tools, like machine learning, thrive on the volumes of data that real-time sensors generate in operations. Early proven examples of edge devices in the upstream involve machine learning algorithms applied to operational work.

4 Edge computing gives frontline managers real forensic information about what is actually going on in the field and do so much closer to real time, and certainly much faster than the usual six-week reporting delay.

5   Operators have already seen real performance gains—more production from existing assets, fewer personnel per operating well, greater leverage for staff, enhanced staff decision-making, and an objective view of the asset that is unfiltered by spreadsheets and human bias.

6   The costs to send humans to the edge to collect data from various old-school sensors and gauges was already high and is now higher because of the costs of carbon and the pressures from the pandemic.

There is almost no limit to the potential impacts that edge could have on the million or more oil and gas wells in North America alone.

Maturing the Edge

To become more mainstream, edge computing and edge devices need to overcome some of the typical concerns from the field.

Telecoms reach: Consumer edge devices like phones and watches are designed to compute away happily while not on a network, but, eventually, they all need to connect up with the home office. They need security updates, new features, and operating system upgrades, all of which entails some kind of network connection. Oil and gas edge devices really prefer always-on, reliable, low-cost networks because these edge devices are controlling dangerous petroleum products. Telecom networks frequently provide limited to no coverage for remote oil and gas fields.

Security: Oil and gas is deeply concerned about the security of its business, since hydrocarbons are so dangerous and the industry is under constant cyberattack. Edge has to be great at security.

Edge monitoring: Keeping a watchful eye on what happens at the edge is both necessary and unresolved. Is there an edge device that monitors the edge computer?

Management of the edge: A proliferation of edge devices will require a new layer of management services to calibrate devices, carry out repairs, and provide installation assistance. What will be the management protocols and practices for providing these services? Is there an emerging services industry that provides edge computer support?

Capital: The industry is capital-constrained, despite the edge advantages of higher production, lower costs, and optimized emissions. This could create the demand for new capital-light business models wherein a third party finances the edge computing in exchange for a share of improved performance.

Architecture: The absence of an industry-level architecture for edge devices is forcing all edge computing services to provide all the layers of the edge stack—telecoms, **bus**es, power, security, apps, sensors, data structure, dashboards, user interfaces, support tools. Provisioning includes operating system updates, app updates, rollbacks, backups, unit testing, and security validation. This usually results in the kinds of walled gardens that made SCADA historically costly. Open architecture may accelerate edge proliferation at lower cost.

Culture: Edge computing has the air of job displacement about it. Overcoming cultural barriers to adoption must be placed high on the agenda for enlightened managers intent on improving the business through edge computing.

Artificial Intelligence and Machine Learning

The landscape of industrial sensors generates so much data that new tools are required to process the volumes, which gives rise to the role of artificial intelligence, machine learning, and advanced analytics in the framework.

Data is fed into AI platforms or programs, which then process and learn from these inputs. Enriched with more and better data, AI can perform more complex and varied tasks, faster and with more accuracy than human analysis.

The maturity and value for AI in oil and gas is huge. It is one of the few industries naturally predisposed to AI because of this wealth of data. AI is faster, doesn't need sleep, and is capable of much of the brute analysis tasks carried out by professionals in the industry. That said, "There is a limit to [AI]. And it needs to be augmented with domain expertise. Engineers are not redundant," according to President of Strategy and Development Azad Hessamodini, from Wood.

The many flavors of artificial intelligence and machine learning provide pathways to process different forms of data and achieve different goals.

AI HAS a long history, originating back to the origin of computers themselves. Alan Turing's role in code-breaking in the Second World War was the catalyst for some of his seminal work on machine learning.[4] The history of the development of computers and AI are closely aligned. While the personal computer was being made by Microsoft in the 1980s and '90s, AI research was ongoing with networked intelligence. More recent developments are a result of improved compute horsepower and agile **DevOps**.

Natural Language Processing

Natural language processing (**NLP**) is AI that can "read" text and extract useful information. Humans tend to use a handful of predictable patterns of wording. Most people don't use more than a few thousand unique words, and so it isn't terribly difficult for AI to be able to read and understand our written languages.

In the business and legal world, this is even more straightforward. Legal documents, white papers, and contracts all use boilerplate text and particular forms of wording for each document and industry. These common patterns are easy for AI to process, read, and interpret.

There are many examples of this kind of AI being applied in oil and gas. Some companies are using this capability to read and analyze legal documents without human involvement. Others are using AI to find critical information in knowledge databases. And others are interpreting engineering content to help write clearer requirements.

Visual Analytics

Machine interpretation of visual data is very mature. Airlines rely on machine readable luggage tags to route passengers' belongings through the maze of conveyor belts that connect baggage to flights. Modern cash registers scan retail tags and figure out what is being purchased, its weight, its price, and any applicable discounts. Even a smartphone can instantly recognize QR codes and act.

These examples all rely on some kind of human-invented specialized symbology—barcodes, bag tags, and QR codes. That's useful but also limiting in situations where there is no practical way to use these symbols.

It should be no surprise that visual data capture and interpretation systems are evolving rapidly to create a clever new category of analytics that works with visual data. In oil and gas, these systems can be "taught" to recognize virtually anything, from intruders at the gate to contractors on the site or wildlife at the fence, even the presence or absence of required safety gear.

And sensors are not limited to data from the visible light spectrum. The software can detect plumes of invisible vapors (like an escaping gas, or steam jetting out from a pinhole in a pipe) and determine the composition of the vapor. This is handy in a world that needs to worry about emissions.

Such robotic eyes are more reliable than human eyes too. Operators need bathroom breaks, take vacations, and require training and supervision, and humans become easily bored with watching screens that don't change frequently.

Visual analytics simplify control room operations, improve oversight of the operating environment, raise compliance, allow for consolidation of control facilities, reduce the drive-around requirements to get eyes on-site, and reduce the operations headcount. This adds up to a safer, lower-cost, more productive operation.

Field service management is a major beneficiary of visual data interpretation. Imagine a supervisory engineer who has contracted for services to a remote asset. Using visual analytics, the system could support the supervisory engineer by automatically opening and closing gates, logging arrival and departure times, and monitoring site activities, inventory moves, fluid levels, spills, and vapors. This moves supervisory engineers out of the field and provides them with greater leverage (i.e., monitoring more services at more wells in parallel).

Satellite and remote imagery offer yet more analytics opportunities. Satellites generate terabytes of visuals but outstrip human capacity to look through all the images to find the one or two that require more thoughtful consideration.

ENTREPRENEURS ARE building robots that can fold laundry.[5] It's devilishly hard to fold clothes. Robots need to distinguish between trousers and tees, fronts from backs, widths from lengths, bell bottoms from capris. Humans do it with ease, but robots struggle. So how do you teach a robot to fold laundry? One way is to have an experienced laundry folder operate a set of controls that manipulates a robot's arms, and by repeating the same movements over and over, with different clothing, the robot eventually learns the job. Human-led machine learning.

Autonomy and Robots

Most AI algorithms are simply learning programs: they only become autonomous when provided with agency. A flying drone inspecting a tunnel is a good example. Once inside a tunnel, the drone must be able to self-navigate. Using sensors that feed it data about the real environment, the drone takes action and executes the "best possible" flying route. That is autonomy; actions enabled by a "neural network" of algorithms and programs.

Digital advances are making it possible to create autonomous versions of a huge range of mechanical contraptions, or at least allow for their remote control without having a human physically on board.

These advances include complex mathematics, learning systems, sensors, data networks, cameras, **robotics**, and digital controllers that have fallen in price and expanded in capability to bring autonomous control within the grasp of most manufacturers and, increasingly, resource companies.

The Extraction Industry and Bots

I admire Rio Tinto's robot strategy. They operate some of the world's biggest mines in challenging places, extracting a broad assortment of rocks, ores, minerals, and metals. They lead the deployment of the supersized autonomous **heavy hauler** in their mines, and this digital innovation is migrating to mines around the world.[6] These massive robots operate without humans at the helm, including in the **oil sands** industry. Miners see 20 to 40 percent improvements in costs and productivity through these technologies.[7] Robots are now being deployed to drill blast holes, conduct aerial mine site surveillance, inspect subsurface and abandoned mines, and search for deposits.

And Rio doesn't stop there. They operate the world's first fully autonomous railroad, in far-off Western Australia, which carries mined material from the inland mine to the port, where the ore is off-loaded to await ships to carry it to market.[8]

There also exist fully autonomous ports that can handle the loading of ships without humans manning the controls.[9] Ghost ships (robot ocean-going vessels) are planned to ply the global trade routes.[10] These innovations report performance improvements between 20 and 70 percent.[11] Australia's resource export ports are likely candidates to achieve partial to full autonomous commodity handling in the years ahead, with eventual integration with the next generation of ships that are themselves fully autonomous.

Rio's resource extraction model has the air of an auto plant, with industrial robots handling the work from end to end. Increasingly, humans don't want to work in hot, dangerous, and demanding locations, and the few willing to do this work want a lot of money for their efforts. Rather than making work safer for humans, mining is simply changing the work to remove humans entirely from harm's way.

Oil and Gas Lags

Compare the mining world to oil and gas companies and their suppliers. After they recruit talent, companies then spend a small fortune trying to keep their human charges from hurting themselves. The amount of time and money being spent on improving safety, training for safety, and measuring safety is staggering. There are legions of safety supervisors, sellers of safety equipment, standards bodies for safety. It's even said in the industry that the way to get a project funded is to claim that it will improve safety or reliability, and as long as the argument is plausible, it gets a look.

It's like we're in a safety arms race that no one can win and no one wants to pay for (try tacking 5 percent onto your bill for your safety program and getting your customer to pay for it). The industry is captive to a narrative that keeps this race going at full tilt.

The next generation workforce, who are much more digitally savvy, will struggle to understand why, in an era when robots make food, fly about, drive around, conduct surgery, patrol oceans, and fold laundry, we still have backhoes with padded furniture. They will spare no thought to the argument that robots take jobs away. Their experience will be that robots actually create better jobs for people.

Getting started with robots is not hard.

Entry-Level Bots

The gateway bot technology is called robotic process automation (**RPA**).[12] A bot executes by copying keystrokes, mouse-clicks, and window-navigation from a "recording" button, performing the task, optimizing itself, and repeating the task while making efficiency improvements each time. They can self-write code, execute that code, and, finally, optimize the code over time.

RPA FIRST appeared in a serious way in the world of online computer video games, as far back as 2001. A "computer-controlled player" is a recorded sequence of repeated keystrokes and mouse movements that executes mundane tasks like collecting weapons and armor, suiting up, and setting game switches. The bots play back at much higher speeds than a human can, and they are so good at that job that they were banned for giving unfair advantage to the human players who built them.[13]

Bot productivity improvements on the order of 80 to 90 percent (versus a planned gain of 60 percent) are common. The London School of Economics reviewed the landscape of published RPA case studies and concluded that ROI ranges from 30 to 200 percent.[14] Up to 75 percent of what we think of as high-value work is actually routine and mundane, and amenable to RPA. Indeed, as John Pillay argues, "the safest bet a CFO can make is on robotic process automation."

RPA bots are built from two digital technologies—a machine learning platform called **CAP**, and a unified test management system, or **UTMS**. The CAP is fed data from some source (text, video images, forms, screenshots, whatever), restructures the data, and inputs into the UTMS, which applies an algorithm to it. Setting up a bot is quicker than implementing other systems because configuration is much faster.

RPA in Oil and Gas

Here are a few areas where RPA should be able to make an impact.

Finance: Financial management in some oil and gas companies consumes as much as 10 percent of headcount. Invoice processing in accounts payable, tax calculations, capturing field tickets, processing royalties and payments, production accounting, financial close processes, and report preparation would benefit from RPA. Some sites have reported that RPA has reduced the time it takes to process an invoice from twenty minutes to forty-five seconds. Cory Bergh contends that "you can automate just about everything in finance with three solutions: RPA, electronic workflow, and business intelligence reporting."

Human resources: Oil and gas companies are not like banks with thousands of employees, but larger producers will have enough employees to require volume-driven administrative roles in HR.

Processing leave requests, handling inbound job applications, updating payroll records, and general HR reporting would be candidates. The shared services centers used by big international oil companies all rely on bots.

Operations: Depending on the company, there will be volumes of repetitive work, moving data between operational systems like SCADA and Excel spreadsheets, and corporate systems recording volumetric data about products, waste outputs, emissions, water handling, energy consumption, and the like. As operations adds more sensors (see the earlier section on the industrial Internet of Things), the additional burden from processing all the data will drive more data capture and analysis, which could benefit from RPA. Other candidate areas are exploration analytics, drilling optimization, compliance reporting, and operational reporting.

The supply chain: Contracting and procurement functions are still very much in control of oil and gas spending to help keep costs from rising, but at the price of introducing more processes. RPA could help with purchase order generation and processing, spend approvals, vendor management, catalogue management, inventory record keeping, and reporting.

The Outlook for Robots

Robots are now commonplace in manufacturing settings where the range of motion is simplified, the level of flexibility is lower, and the environment is tightly controlled. Similarly, subsurface, virtual, and aerial robots are already established in oil and gas, and I expect to see the use of virtual robots growing significantly for many years. The big open frontier is all of the terrestrial activity that could be targeted, but until autonomous tools become far more adept at navigating the terrestrial landscape, they pose risks that many oil and gas companies consider unacceptable.

Cloud Computing

In some quarters of the oil and gas industry, cloud computing is no longer even a discussion. It's assumed that cloud computing is the platform of choice, is already a key part of infrastructure, or is a minimum condition. According to Azad Hessamodini, "cloud is not a thing anymore. It's a given. You can't do machine learning without cloud."

Realized Cloud Benefits

The benefits of cloud are now thoroughly ahead of the costs.

Improved agility: Massive cloud data centers reliably deliver **compute load** and data-storage capacity in almost unlimited amounts nearly instantaneously for whatever computational problem might be at hand. Need to crunch a dataset measured in multiple terabytes? No problem. Want to run multiple versions of a complex model (such as a refinery LP) simultaneously? Easy. Need to flash assemble a highly scalable system to manage through a conflagration? Just a call away. No more issuing requests for capital to expand the data center, or waiting months and months for servers to show up from manufacturing.

Better security: Concentrating data and computer capacity into massive facilities must surely be enticing to cybercriminals. But it turns out that these dense facilities are more efficient at security management. The instant a mitigation action is deployed in response to a threat, that action can be propagated immediately to the other facilities in the network. Vulnerabilities can be more quickly identified and remediated. Patch management is simplified.

Greater efficiency: Big cloud service centers are more efficient. Compute loads can move around the facility in response to weather conditions, as Google found out when they subjected their energy profile in their data centers to AI analysis.[15] Moving the loads to the shady side of the building (reducing air conditioning needs), meant that Google could save 40 percent of its energy bill. Cloud operators design their own servers and farms to wring out cost, along with their own management software, backup and recovery solutions, and plant optimization. Lightly used gear can be quickly deployed to meet needs.

Cost effectiveness: Cloud offerings are not always a cheaper option than running your own data center, particularly if you're already at scale. The main costs of a data center are electricity and manpower, and, depending on where you locate your facility, you can possibly take advantage of power pricing and lower labor costs. However, you do sacrifice access to capacity on demand, your responsiveness is lower, and you own the security problem. And you are likely only as efficient as the management software you purchase enables you to be.

The Great Enabler

Cloud computing is the key enabler for so much innovation. Consider the edge device. While the simplest edge computing structure is a single isolated edge device running quietly, eventually the device needs to be updated (perhaps frequently, given the need to keep current with security patches). In the same way that your smartphone receives updated apps from your favorite service, edge devices receive updates from the cloud.

While a single edge device will certainly add value, it's when the data from multiple edge devices are combined and analyzed that machine learning becomes very powerful. Cloud is the platform to aggregate the data, house the algorithms, and provide the analytics.

Early in their maturity, cloud services were mostly a commercial IT question, but that is no longer accurate. As one of the case companies illustrates, cloud computing also works for operations.

Not Cloudy Everywhere

The advantages of cloud computing may not accrue equally to all oil and gas companies and countries:

- Cost benefits in cloud accrue from aggregating many small pockets of demand into very large facilities. This formula doesn't work where the demand is already highly concentrated.

- National oil companies may be concerned about holding nationally sensitive data about oil company operations in cloud data centers.

- Some national oil companies will be less sensitive to competitive pressures, carbon sensitivity, and capital market scrutiny.

Cloud computing is not always the answer in all situations, but it is now the default starting position for architectural choices in a digital world.

It seems inevitable that cloud services will continue their relentless march into oil and gas. The costs and capabilities of cloud are already compelling, the software companies whose products run the industry are assuming cloud deployments in the future, and most of the innovations coming to the industry are based on cloud availability.

One very large attractive future market includes the operating technologies or SCADA environments that often run separately from commercial IT landscapes but are also becoming more digital and moving more to the cloud. One of the case companies has achieved exactly this future.

Blockchain

Distributed ledger technology, also called blockchain, is a database design first described in 2008 as a way to maintain records between parties who don't trust each other, but without a central agent to provide trust. A blockchain database has transactions (records) that are recorded sequentially by time stamp. The blockchain database is copied to many different computers, which use complex encryption and mathematics to update the database (by consensus) and ensure that the transactions cannot be tampered with.

Bitcoin is the high visibility blockchain use case that has captured the imagination of many investors, but blockchain has so many other uses that can create value in oil and gas by lowering costs, eliminating disputes, and detecting fraud. Key areas include supply chain management, **track and trace**, trade simplification, automating contracts, and managing carbon credits.

Blockchain Simplified

Imagine the classic buy/sell relationship that we are all familiar with. I sell a book (which I do), and you wish to buy the book (which you do). We agree on the unit price, the number of copies you want, the delivery terms (shipping address, shipping method), and the delivery date. Being a simple fellow, I keep track of these details on a spreadsheet (or a ledger), which includes your name and address, the number of copies you want, and all these other details.

Since your memory may be faulty, or perhaps because you don't quite trust me, you write down the same details on your own spreadsheet or ledger. Now we're both maintaining the same data, but in separate ledgers. Either of us could have written our side of the transaction incorrectly, which could give rise to some confusion later—was it ten copies or twelve?

We rely on many central authorities to keep track of some key ledgers. Banks manage money ledgers. Governments look after tax ledgers. Stock exchanges keep track of trades.

Now let's string a few steps together. After you place your book order, I place an order with Amazon to print the copies, and Amazon places orders with a trucking firm, air transport, and ground courier to finally deliver the books to you. Each of these parties has their own ledger to keep track of their bit of the deal. And each ledger has the potential to contain errors or differences with the other ledgers.

If all of these ledgers could be precisely aligned all the time, eliminating the possibility of confusion and error, without need of a big central authority like a bank, we could eliminate much of the cost we incur to deal with the inevitable errors and confusion.

That's what blockchain does—it creates highly trustworthy data without the need of a big costly and vulnerable central authority. Transactions of interest are grouped together into blocks, encrypted, and digitally attached to each other in a large chain. The chain sits on many computers simultaneously, and the chains all have to match.

Such a structure is highly secure because corrupting it means corrupting the majority of the chains at the same time.

Blockchain has a lot of potential in the industry, especially in the back office. According to Cory Bergh, "a lot of effort in the back office in oil and gas is to reconcile why there are differences between the information in different systems. Consensualized data solves this. Blockchain is the perfect tool for consensualizing data."

Bulk Commodity Transfers

Water hauling is a big issue for the North American upstream industry. Water is used for fracking, steam generation, and water flooding. Water is produced naturally from wells and flows back up after fracking and steaming are complete. Contaminated saltwater needs to be disposed of at special sites, and every ounce must be accounted for. Local governments monitor water resources carefully but are resource constrained. Invoicing and reconciliation are largely manual, and compliance is self-reporting. There are well over 100 million truck movements per year hauling water in North America alone. Regulators want precise, frequent, and accurate records of water handling and levy penalties or production curtailment for repeat rule-breakers.

GumboNet is a blockchain-based solution to simplify and streamline this costly and inefficient process. Eventually, other bulk commodities, including chemicals, sand, and fuel, could benefit. Haulage of equipment may also be improved using blockchain.

Trading Simplification

One of the earliest trials of blockchain was in the area of petroleum trading, which incorporates considerable haulage. Trading in petroleum (gasoline, diesel, jet fuel) is tricky because the product is highly regulated, very valuable, completely standardized, quality sensitive, easily blended, **fungible**, and bulky. It passes through many hands (tank farms, barges, tankers, pipelines) and title changes frequently mid-journey. It crosses borders easily, attracts lots of taxation and tariffs, and generates its own wave of paperwork.

VAKT is a blockchain-enabled reimagining of the trade relationship between parties in the industry.

Product Certification

To assure product quality, oil refiners operate labs that process thousands of assays and samples annually to confirm that oil purchases and refined products meet regulatory and industry specifications and to track those samples as they transit to customers. Labeling errors, information losses, misplaced samples, or samples disconnected from their documentation frequently happen, triggering considerable work chasing down samples.

Repsol has launched a blockchain solution called BlockLabs to transform this process area.[16] BlockLabs is built on a blockchain middleware platform called MARCO from Finboot, a technology company based in London, Cardiff, and Barcelona.

AFE Balloting

Authorization for expenditure (AFE) is the process to approve capital spending for a project, such as a well or an infrastructure investment. In the case where a project has multiple possible participants, the process to secure agreement between the parties is called balloting.

The lead operator in a joint operating agreement (JOA) prepares a draft AFE ballot to send to the other participants. The draft AFE sets out the proposed spend for each of the participants, pro rata to each one's interest level. Ballots are binding and take the form of a certified, hand-signed, manually delivered letter. The process is time-consuming, paper-based, costly, manual, and prone to disputes.

GuildOne is addressing this problem area working with Blockchain for Energy, a consortium of energy companies based in the US.[17]

A Blocky Future

Blockchain's future looks assured. The early use cases are demonstrating quite dramatic cost savings, and the solution areas are almost certainly the easiest ones to prosecute at this stage. The big opportunity areas have yet to be tapped.

The low-code, no-code technology philosophy will also come to the blockchain protocols, which will simplify and democratize its adoption cycle. Next generation solutions will be very easy to use and operable on all platforms, further insulating the analyst from having to become a deep blockchain expert.

Eventually, there will be a wide diversity of blockchain database designs, each solving for specific kinds of problems. Some will focus on the high volume but low data–intense applications, such as bitcoin. Other companies will aim at **smart contract** opportunities. And governments will embrace different blockchain databases for registries, contracting, and compliance needs. Consequently, companies should anticipate that they will likely be party to multiple blockchain solutions, and that attempts to rigidly standardize blockchain protocols are unlikely to succeed.

While blockchain solutions are generally economic on their own, their real impact is realized in combination with other technologies. For example, using blockchain to immutably record machine-generated data from an edge device unlocks trusted autonomy for that edge device and its recorded data. The industrial world seeks precisely this kind of advancement in the drive to decarbonize distributed carbon sources and carbon sinks.

The proliferation of blockchain-enabled solutions will create demands for interoperability between various blockchain structures. Solution providers will need to incorporate open standards and APIs into their designs.

Enterprise Systems

The final layer in the digital foundation is enterprise systems, which includes three different classes of digital solutions that are pervasive in the industry:

- cybersecurity
- platforms
- applications

Cybersecurity

Cyber criminals are benefiting from digital advances for the same reasons that oil and gas is gaining. Much of the digital world is based on open-source technologies, a kind of rocket fuel for propelling innovations forward, including those "innovations" with hostile intent. Digital solutions are very democratic—anyone can access them. The apps, including software code for viruses, can be freely or cheaply had, which tells you they're not costly to make. The computer coding languages are pretty easy to learn, and the techniques for making the apps seductive and mildly addictive are widely shared. One of my university courses almost forty years ago instructed students in how to make self-replicating software, a key feature of computer viruses.

These commonplace digital technologies from the industrial world of energy manufacturing and distribution have found an unwelcome home for the nefarious, namely, cyber activity. The Colonial Pipeline ransomware attack in spring of 2021 showed just how vulnerable energy infrastructure is to cyber threats.

ON MAY 7, 2021, Colonial Pipeline's systems were successfully breached by cyber criminals in a ransomware attack.[18] The pipeline was disabled for several days, causing price spikes, panic buying, and supply shortages along the US east coast. Like many companies, the pipeline operator had a handful of log-on credentials for a virtual private network (VPN) account that allows employees and contractors to access Colonial's systems from remote locations. These log-on IDs were found in a cache of leaked passwords on the **dark web**.

For various reasons, conditions are ripe for much more widespread cyberattacks:

- We have a growing reliance on modern 5G wireless network connectivity, but much of the world still labors under 2G, an older and more vulnerable telecom standard. Wireless links can be compromised at source and during the transmission of data. Both links and transmissions can be hacked.

- We are adding sensors of all stripes to many things, creating a greater attack surface for cyber criminals to target.

- We are interconnecting our systems, which allows for faster spread of viruses and criminal access. The response time available to deal with a threat is shrinking. Computer viruses now spread much faster than human viruses. Staying on top of all the patches is a monumental task with the result that many successful attacks target unpatched kit.

- We are adding internet links to our legacy infrastructure. That legacy gear was never designed for such a hostile world and lacks the ability to be patched, or even monitored for cyber activity.

- Finally, we're unleashing a brand-new wave of innovation—autonomous transportation and drones, smart manufacturing, smart cities, the metaverse, digital farming—that will add to the opportunity for cyber criminality in ways we have yet to fully understand. Forward-thinking criminals are already preparing for this new lucrative playground.

Unfortunately for business, cyber activity pays off, and handsomely too. In the US, thousands of ransomware attacks take place daily, with an average ransom payment over $13,000. The probability of getting caught, and lenient punishments, do not appear to deter the market.

The US Department of Homeland Security tries to keep tabs on cyber activity and believes that over 50 percent of all cyberattacks from 2015 to 2019 were aimed at energy infrastructure (power, oil, and gas), rather than banking. Much cyber activity originates with state actors who have interests in destabilizing entire economies for reasons other than purely theft of financial assets.

The costs to the victim are much higher than any ransom payment, in terms of an urgent and unanticipated outlay to remediate the attack, an unwanted distraction from operations, a shut-down of operations to recover, the costs of brand damage, the potential for customer defection, and potential regulatory penalties. While on a panel with me, experts from RigNet estimated that the average cost to recover from a successful attack in energy is over $17 million, more than five times the average across other industries.

The lightweight human-centric tools of the past for managing cyber activity simply are no longer up to the task of managing and repelling the onslaught of attacks. With thousands of access points, sensors, equipment, networks, and industrial assets, each a potential cyber target, companies need all new tools to deal with the rising volume of activity.

Leading companies approach this new problem by applying the latest digital tools, including AI, machine learning, and robots, to cope with cyber activity. The resulting struggle pits the human ingenuity, AI tools, and bots of the criminal sector against the trained technical teams, AI tools, and bots of industry. The clash is like a cat-and-mouse face-off taking place entirely in the ether, with the cat having to respond to every move the mouse makes.

Given the complexity of the environment, if your company is not already bringing behaviors and digital tools such as the ones listed below to the combat, you're showing up to a fire fight with a butter knife.

- Your board takes an active interest in cyber issues, holds regular education sessions on cyber topics, and has quarterly briefings from security experts on cyber activity.

- Employee education programs incorporate cyber awareness training to highlight the perils of unprotected devices, **phishing** attacks, and **spoofing.** Some campaigns even include fake phishing attacks that help capture inattentive employees.

- Risk-review committees flag cyber risks alongside operational risks as high likelihood and high impact. That way, cyber defense gets some organizational attention.

- Encryption is on by default for everything—data, devices, sensors, and data flows. Since it's only a matter of time before quantum computing compromises digital assets, better that they are at least encrypted.

- Particularly sensitive functions, such as encryption, are handled by hardware that can force hackers to need physical access to carry out an effective attack. Hardware-based encryption also lessens the overhead burden on networks.

- The non-sleeping, always-learning AI and bots actively monitor the digital environment to detect intrusions, isolate intruders, repel attacks, and neutralize many cyber threats.

- Cyber expertise is organizationally separate to bring independence to standards, testing, and monitoring of the digital assets. Services run a continuous program of penetration testing to detect weaknesses to be corrected.

- Access to company digital assets and resources by third parties, suppliers, and contractors is time-boxed.

There are many kinds of arms races in industry. One is the safety arms race, where companies invest feverishly to protect their employees from all possible harm. Another is cyber. I see no inevitable conclusion to this race—it looks well set to run for the very long term, and it's escalating as the prize for the criminals (and the costs to business) ramps up.

Platforms

My first serious corporate job was with Imperial Oil, where I supported a corporate system called CORPS. Big companies love abbreviations—I can't recall what exactly CORPS stood for, but I remember precisely why it existed: to save data center time in mounting and dismounting magnetic tapes. It also turned out to be my first exposure to platform technology.

CORPS was a middleman or middleware system. At the time, in an era before enterprise resource planning (ERP) technologies, Imperial Oil had a handful of major commercial business systems that handled different aspects of product movement, which individually fed data to many other systems. The dozens of individual data feeds from one system to another created a literal Gordian knot of integrations. CORPS was an attempt to solve this problem—all the inputs fed into one gigantic master file, which then generated all the individual data feeds. It was simply faster than running all the data transfers individually.

With hindsight, CORPS solved a many-to-many problem. Many data inputs going to many data outputs creates huge cost, as each data supplier needs to maintain an individual connection with each data consumer. A change to any one system can have a ripple effect on many other systems.

Modern digital platforms are very good at solving the many-to-many problem. For example, trading platforms facilitate buyers and sellers to find one another and transact. Amazon matches many suppliers of goods with many customers. Airbnb matches parties holding available accommodations with travelers needing a short-term place to stay. Uber matches cars and drivers with customers. Platforms often aim to capture network effects by connecting very large numbers of counterparties.

Unsurprisingly, oil and gas has already discovered a few solid opportunities for creating trading-type platforms for its various businesses.

USED AND SURPLUS GOODS

Capital projects in oil and gas frequently purchase technology and equipment that are surplus to actual need. This excess spending creates an inventory of parts and spares that solve for various delivery risks. For example, some items have long lead times, and if the ordered equipment arrives damaged or unserviceable for some reason, the entire project faces a delay awaiting a replacement. Having a second, undamaged item on hand might help the project keep on schedule, but might otherwise result in an end-of-project inventory for disposal.

These inventories are hard to liquidate for the following reasons:

- Buyers want warranties for the equipment they purchase, which sellers may not have.

- Sellers are rarely set up as merchants, with catalogues, photos, and pricing of surplus goods.

- The surplus items may be located in hard-to-reach locations, creating a logistics burden.

- The condition of the surplus goods needs to be checked before sale.

- The specifications of the surplus items need to be accurate, current, and documented.

Facebook pages for disposing of surplus oil field inventory have appeared, along with bona fide digital startups working on the problem.

SERVICES

Another platform example is for services. In particular, upstream oil and gas assets (wells and facilities) consume a huge range of services from thousands of suppliers. The locations of the suppliers are material—too far away, and the mobilization costs and safety risks escalate. The urgency to get services can overcome the commercial need to solicit competitive quotes, resulting in higher than necessary costs. The nature of services specific to individual assets varies dramatically, making price discovery complicated. The lack of visibility into the overall supply and demand for services in a region results in a hugely suboptimal market and high emissions costs.

One answer is to create an internal platform that leverages procurement systems to attempt to optimize spending within a company. Products like Salesforce and ServiceNow fill this need.

Another more powerful solution is the external platform that aggregates the services on offer with the demand and attempts to best optimize the allocation of services to that demand. Eventually, with enough data and enough participants, new analytic tools can bring incremental value to the overall services needs of a region.

INFORMATION

Similar to the CORPS system, platform solutions that solve for the problem of many data sources and many data suppliers have appeared for single company use. These are found principally in upstream oil and gas given the diversity of commercial software packages in use, the wide range of data sources (ERP systems, paper binders, PDFs, **data lakes**, spreadsheets, **historians**) and the numerous unique internal uses for the data (well planning, capital budgeting, geologic analysis, environmental studies). A big difference is that once data from the many disparate sources are harmonized, unified under a single taxonomy, and organized appropriately, creative new workflows and business logic can be programmed to work with the data directly. New ways to visualize the data analytically can help accelerate decision-making.

Notably, these information platforms are becoming available as cloud services instead of on-premises technology, which accelerates adoption, lowers ownership costs, and improves security.

Platforms are already too valuable to ignore, and with the pressures to solve for problems such as carbon tracing, industry-specific platforms will grow in prominence and affordability.

Applications

Enterprise systems, from the likes of SAP, Oracle, and Microsoft, are now ubiquitous across oil and gas and are the true workhorses of the commercial world. Without them, the product does not get delivered, the bills do not get collected, the employees do not get paid, and the financials are not accurate. Intertwined with the behemoths are hundreds of other solutions, each playing to their respective areas of strength. And once installed, they are usually for life, as the costs of transitioning generally outweigh the benefits to be gained.

Unsurprisingly, enterprise systems have rapidly embraced digital technology because these systems are also widely deployed in industries much more impacted by digital technologies than oil and gas.

Cloud enablement: Browser-based versions of ERP technology date back to the early 2000s. PeopleSoft (acquired by Oracle in 2004), was rewritten as a web browser solution in 2000, which positioned it for a cloud future. SAP released a web version in 2004.

Integration and APIs: At one time, the design of enterprise systems was intended to lock customers into their walled gardens of functionality and to keep at bay the unreliable data from outsiders. Today, ERP systems offer API sets to enable better integration with third-party systems. It's not perfect, but it's superior to a custom-developed integration.

Support for things: If harnessed, the world of smart things (edge devices, mobility, apps, sensors) can unlock many new clever commercial solution designs, from instant payment via smartphones, to real-time inventory visibility, to anytime tracing of the supply chain.

Strong security: Their large installed footprint means costly security overhead functions, such as multi-factor authentication, are spread out across a global base of customers. The individual costs of security to specific customers are very low.

Deep analytics: With so much data aggregated in their bowels, enterprise systems unlock the power of analytic services such as AI and machine learning, amplifying impacts.

Advanced features: Capabilities such as blockchain support are built into the ERP systems, which will cut deployment time and costs.

Third-party apps: Much like how Apple created the App Store to allow third-party software developers to create innovative products beyond the set produced by Apple, the ERP vendors are also enabling third-party applications to engage directly with their products.

As I see it, these enterprise solutions become more important in the future, not less. They are exceptional at scale work, their costs on a per-user basis are low, and they are ideally situated for a world with more robots and autonomy that is predicated on high-quality data.

KEY TAKEAWAYS

Digital innovations are evolving rapidly and experiencing different levels of uptake in oil and gas. Here are some of the key takeaways from this brief survey of the digital technology landscape:

1   The Digital Framework—business capacity, the digital core, and the digital foundation—is a useful tool to help grasp why digital embraces technology, process, and people.

2   The digital core—comprising data, the IoT, AI, and autonomy—is infinitely configurable to create signature ways of working.

3   The digital foundation—cloud computing, blockchain, enterprise systems—is table stakes: if you don't have it, you're not in the game.

4   Business capacity—the adoption of agile methods and user experience, and robust change management—is the means to embrace digital from a process perspective.

5   Even the operational systems in oil and gas, from remote devices to traditional control rooms, are benefiting from digital enablement.

6   Data skills are central to the operations and management of oil and gas.

Notes

1   Gordon E. Moore, "Cramming More Components onto Integrated Circuits," *Electronics* 38, no. 8 (1965): 4.

2   Carl Shapiro and Hal R. Varian, *Information Rules: A Strategic Guide to the Network Economy* (Boston: Harvard Business Review Press, 1998).

3   Robert U. Ayres, "Barriers and Breakthroughs: An 'Expanding Frontiers' Model of the Technology-Industry Life Cycle," *Technovation* 7, no. 2 (May 1988): 87–115, doi.org/10.1016/0166-4972(88)90041-7.

4   Andrew Hodges, *Alan Turing: The Enigma* (London: Burnett Books, 1983).

5   Evan Ackerman, "Is There a Future for Laundry-Folding Robots?" IEEE Spectrum, April 29, 2019, spectrum.ieee.org/automaton/robotics/home-robots/is-there-a-future-for-laundry-folding-robots.

6   "Rio Tinto's Autonomous Haul Trucks Achieve One Billion Tonne Milestone," *Rio Tinto*, January 30, 2018, Rio Tinto, riotinto.com/news/releases/AHS-one-billion-tonne-milestone; Deborah Jaremko, "Canadian Natural Planning Test of Autonomous Oilsands Heavy Haulers," JWN Energy, March 1, 2018, jwnenergy.com/article/2018/3/1/canadian-natural-planning-test-autonomous-oilsands.

7   Kevin Smith, "Rise of the Machines: Rio Tinto Breaks New Ground with AutoHaul," *International Railway Journal*, August 9, 2019, railjournal.com/in_depth/rise-machines-rio-tinto-autohaul.

8   Kevin Smith, "Rise of the Machines."

9   Clement Ruel, "Autonomous Shipping Ports," Global Infrastructure Hub, December 9, 2020, cdn.gihub.org/umbraco/media/3596/42-autonomous-shipping-ports.pdf.

10   Nick Savvides, "Revolution for Inland Shipping Depends on the Success of the Yara Birkeland," *FreightWaves*, June 6, 2019, freightwaves.com/news/revolution-for-inland-shipping-depends-on-the-success-of-the-yara-birkeland.

11   Ruel, "Autonomous Shipping Ports."

12   Clint Boulton, "RPA Is Poised for a Big Business Break-out," *CIO*, June 12, 2019, cio.com/article/3269442/software/rpa-is-poised-for-a-big-business-break-out.html.

13   Wesley Yin-Poole, "Blizzard Bans 74,000 World of Warcraft Classic Botters," *Eurogamer*, June 18, 2020, eurogamer.net/articles/2020-06-18-blizzard-bans-74-000-world-of-warcraft-classic-botters.

14   Mary Lacity, Leslie Willcocks, and Andrew Craig, "Robotic Process Automation: Mature Capabilities in the Energy Sector," The Outsourcing Unit Working Research Paper Series, London School of Economics, October 2015, 19.

15   Richard Evans and Jim Gao, "DeepMind AI Reduces Google Data Centre Cooling Bill by 40%," DeepMind, July 20, 2016, deepmind.com/blog/article/deepmind-ai-reduces-google-data-centre-cooling-bill-40.

16 "Blockchain Technology for the Energy Sector," Repsol, 2020, finboot.com/post/blockchain-technology-for-the-energy-sector.

17 GuildOne, "GuildOne's Royalty Ledger Settles First Royalty Contract on R3's Corda Blockchain Platform," GlobeNewswire, February 14, 2018, globenewswire.com/news-release/2018/02/14/1348236/0/en/GuildOne-s-Royalty-Ledger-settles-first-royalty-contract-on-R3-s-Corda-blockchain-platform.html.

18 Raphael Satter, "Colonial Pipeline: What We Know and What We Don't about the Cyberattack," Reuters, May 10, 2021, globalnews.ca/news/7848118/colonial-pipeline-cyberattack-what-we-know.

BUSINESS MODEL
TRANSFORMATIONS

"What you don't want is a new asset to quickly become an old asset simply because you designed it in an old way."

CORY BERGH, VP, NAL Resources

IGITAL INNOVATIONS can be trivial (my Apple Watch reminds me to wash my hands for twenty seconds) as well as profound (I abandon in-person retail shopping for online ordering). The stay-awake worry for boards and executives is the risk of an entirely new business model taking shape unnoticed and with sudden and dramatic arrival, upending a long-established business, its employees, and its shareholders. This chapter considers how the industry is being reshaped by new business concepts and models that are made accessible by digital technologies.

Upending Successful Models

I vividly remember my first Uber experience. I was in Perth, Australia, on business with a colleague. We were just wrapping up an hour-long whiteboard session preparing for a client meeting.

"Shall I call a cab?" I asked.

"No worries, mate, I've already booked an Ubah," he replied.

How was this possible when we had only just finished our discussion? He'd had no opportunity to phone. Minutes later, we boarded an impossibly clean, unmarked vehicle, and we set out without telling the driver where we were going. We exchanged a few words with the driver about the weather and cricket, and when he dropped us off, we left without paying.

Every assumption I had about a taxi service was shattered, from how you book it to how you pay. As a business model, Uber has been profoundly impactful on the taxi sector. The first recorded taxi service dates back to the 1600s, and the business model of a conveyance for hire for short trips had not fundamentally changed since its inception in London.

I once naively treated the modern oil and gas industry as having just one business model: explore for resources, economically extract oil and perhaps gas, refine it into valuable products, and distribute it to a constantly growing global market. The integrated international and national oil companies have perfected this model over the past hundred years, to the point where our energy resources are more plentiful and less costly than drinking water.

Hidden in this construct are myriad lesser business models. Refining and gas processing are continuous manufacturing businesses. Greenfield facilities start out as very large and complex construction projects. The world's largest retail chain belongs to an oil company. The sums of money involved in oil and gas can approach that of a sizeable bank. Arranging for services from a hugely fragmented supply chain is similar to organizing a movie production. Distributing the finished products to every nook and cranny of the planet is an exercise in managing very complex logistics.

These embedded and not entirely obvious business models have been very stable for the better part of twenty years, and in many cases, much longer. Oil refineries run largely uninterrupted for thirty years and mostly adhere to their original designs. Other than retail assets, oil and gas facilities are rarely put through a complete makeover. Inevitably, the pathway is to cling to the original design as long as possible. When the economics erode, assets are sold off, wells are abandoned, and refineries are converted into storage sites.

Fresh perspectives can often bring into focus how business models might change. For example, to a Toyota production engineer, drilling a sequence of gas wells looks suspiciously like a reverse auto assembly line, where the workers and their kit move about and the car stays in one place. To a Hollywood studio, distributing subsurface data beyond the walls of a resource owner bears a strong resemblance to movie streaming. To an organic produce farmer, if it's possible to manufacture an organic tomato sauce, why not a branded sustainable diesel fuel for the farm?

I HAVE indirectly experienced a number of business model changes impacting the oil and gas industry over the years, but almost exclusively, these changes have been driven by resource development. Canada has shifted its oil industry decisively in favor of its enormous oil sands resource. The combination of **hydraulic stimulation**, or fracking, and horizontal drilling, unlocked monumental sources of gas in the US. Australia became the world's largest gas exporter in part by capitalizing on its proximity to Asia and by figuring out how to monetize its gassy **coal measure**s. What's been missing are the non-resource transformative business models, such as Airbnb, Netflix, Uber, Google, and Tesla, that have impacted so many other markets.

One of the techniques used to mount an effective challenge to an established business model is to identify the underlying rules, norms, and assumptions by which the established players operate, and by finding ways to creatively break those rules.

Hundred-Year-Old Orthodoxies

Imagine yourself eons ago, when life was more rugged and dangers were everywhere. Out gathering kindling for a fire, you suddenly encounter some wild animal. Your survival depends on the primitive part of your brain, the amygdala, to help you to detect and react instantly to this situation: it floods your body with adrenaline, boosts your heart rate, and triggers higher order processes to help you decide whether to fight the threat or, more likely, take flight.

This most certainly makes you, the early human, a jittery and jumpy beast, in a perpetual state of near panic. It takes a lot of energy to be on high alert all the time, so eventually, as a learning creature, you figure out that certain settings, like dark caves or tall grass, sometimes harbor hidden dangers. You learn to avoid these potential threats, helpfully lower your guard, and burn less energy just staying alive. Most importantly, you teach these same life-saving rules to your offspring.

These kinds of rules are the orthodoxies of life—principles or guidelines that everyone believes in, and no one really questions or challenges because they are generally reliable. For example, when my children were growing up, I carefully taught them not to talk to strangers or get into a stranger's car (the tall grass of our day). Underlying this rule was my fear that they might be abducted. I might try to explain the logic behind the rule to my child, but how can you expect a six-year-old to grasp the meaning of a kidnapping? Did they really get it?

And now, using Lyft, my kids call up complete strangers and get in their cars.

Just consider all the dimensions of the taxi industry, and how the ride-sharing business model upended all of its cherished rules to reinvent an entire sector in just a few years.

Business Model Feature	Traditional Taxi Service	Ride Sharing Service
Capacity model	Capped capacity	Surge capacity
Financing	Company owns car	Driver owns car
Localization	City-specific phone numbers	Globally consistent
Resource allocation	Human dispatcher	System dispatcher
Operations	Driver selects routing; uncertain arrival time	System selects routing; predicted arrival time
Performance management	No feedback loops	Two-way feedback
Technology	Call centers, radio, in-car payment	Devices, cloud, AI

Many commercial business models have been upended by digital innovation:

- Hotels are big and expensive buildings financed by companies for overnight stays. Then along came Airbnb, which converted millions of unused rooms in homes around the globe into rentable space. Airbnb is now the largest hotel chain and doesn't own any hotels.

- Music distribution required the logistical skills to handle the distribution of CDs. Then along came Napster and Apple Music, which allowed customers to acquire just one song at a time and create their own playlists composed of just the music they like.

- A limited selection of books were either sold in bookstores or borrowed from libraries. Then along came Amazon, which carried unlimited titles with fast delivery to the home or office, and Kindle, which revolutionized the distribution model of digital books.

- Movie rentals meant a late-night visit to Blockbuster in a vain hunt for something to watch on VHS or DVD. Then along came Netflix, which offered access to an unlimited number of titles, on demand and available instantly, for a monthly subscription fee.

- Getting the forecast meant tuning in to the local weather network, often after a long morning news broadcast. Then along came Google Home, which tells you the forecast when you ask.

- Entertaining your family involved gathering everyone together in one room to watch the TV, with the programming often dictated by the networks. Then along came VPN, Wi-Fi, and tablets, letting individual family members watch what they want, when they want it, from anywhere in the world, commercial free.

The pressures to decarbonize, the shortage of capital, and the possibilities presented by digital innovations now expose many of the hundred-year-old embedded rules and assumptions in oil and gas as candidates for a rethink.

Orthodoxy 1: Data Has No Value

The first rule or assumption that is tumbling away, but not yet uniformly, is that data in oil and gas is of secondary value and is not an asset on par with physical plants, equipment, and resources.

The industry believes that all of its data is proprietary, must remain inside a firewall, and must be heavily protected. Data is recorded as an operating cost, which minimizes the capital allocated to it. At one time, collecting and storing data was very costly, and that cost created a barrier.

Today, data struggles against these constraints. It's so cheap and easy to generate, collect, store, distribute, and replicate data, that data now wants to be free. Azad Hessamodini put it best this way: "The understanding of 'data' is the least mature part of 'digital' in oil and gas. The liberalization of data is just not happening. It's a shame."

As an industry, oil and gas is blessed with enormous holdings of data, and it generates copious quantities every hour, but it can't begin to analyze it all anymore, and it's missing out by clinging to the idea that only industry insiders can make sense of it.

Digital's first movers in oil and gas see the profound importance that data holds for their future. They recognize that the business models of the industry have been largely perfected in their current state. Waiting for some chemistry or physics breakthrough is folly.

Data is now an asset and has value, but where exactly?

Orthodoxy 2: Humans Do the Work

I learned a hard rule at my first oil and gas job, where I was told I could not be promoted because I wasn't an engineer. I couldn't be trusted to supervise any equipment since I didn't know how it worked. For a long time I accepted this rule, and many still do.

Oil and gas managers generally believe that the work of the industry is complex and cannot be automated. Work requires high levels of skill, years of training, and human intelligence to execute. Only engineers can engineer; only geologists can combine the art and science of interpretation.

Even junior jobs are still very manual. Oil and gas still features an abundance of jobs that require humans to endlessly drive around, stare at dial readings, take paper notes, fill in spreadsheets, and phone-in or email findings.

Jobs are built around the **dumb metal** in the industry: the constraints, operating parameters, and safety needs of the metal. Few installed assets feature modern sensors, computational smarts, and communications support. Adding sensors to operating devices like pumps is usually deemed to be too costly because of the MOC process. And, frankly, the existing kit still works. It's not exactly broken.

But the cost of new sensors has been tumbling, and it will continue to fall until it approaches zero. Newer digital designs for operating equipment require much less physical intervention. Sensors can simply be strapped or magnetically attached, and then can tap into vibrations, heat levels, and noise outputs. Sensors will soon be disposable, to be flushed into pipelines as smart balls (instead of PIGs), or part of a drill bit, or bolted to the side of a pump.

The one dimension of digital technology that has yet to make a meaningful appearance in oil and gas is the transformed user experience. Most technology providers are not yet aiming for frictionless, rapid rollouts and zero-training launches. The nifty inventions from the consumer world (likes, gamification, endorsements, badges, rewards, stars, feedback, rankings) stand to make a huge contribution to the cost and quality challenges of the industry.

Someday, job definition will start with the required level of automation and digital support, and then define the role of the human.

Orthodoxy 3: It Takes a Team

Along with the notion that only humans can do the work of the industry is that work is best done in teams working closely together. This rule abruptly tumbled away during the pandemic; oil and gas professionals, particularly those involved in intellectual work and not tied directly to a physical asset, are very capable of carrying out their jobs from just about anywhere.

Of course, physical plant workers will still need to be in proximity to the plant, and much of their work requires more than one worker for safety reasons. But the office towers in many big oil towns—partially vacant because of the oil market oversupply of 2014, hammered in the market collapse of 2020, and now forced shut in the pandemic—may not reopen.[1]

Teleworking has been around for a decade or more but has never really taken off, because it requires a tipping point. In oil and gas, with its top-down, command-and-control, siloed organization structures, hard-wired with budget systems, it would take but one or two powerful people in the organization to insist that teleworking is "unsuitable," and that would be that. Now, the orthodoxy has been shown to be incorrect. Teleworking seems to function fine when everyone does it.

A few very specific roles in oil and gas have their own dedicated office facilities. Oil traders in particular have operated out of purpose-built trading floors. During the pandemic, traders moved to work from home. Yet oil and gas products are still being successfully traded, financed, and delivered. Such dedicated trading facilities may be endangered.[2]

Face-to-face collaboration sessions with key suppliers using white-boards for problem-solving have also ended, despite the constant need to work on the problems of the industry. New equipment doesn't work properly. Business systems have to adapt to new rules. Ongoing compliance projects need to hit their deadlines. Problems are still being solved.

Selling to oil and gas has favored a business-to-business approach. Sales has always involved lots of in-person meetings, coffee discussions, site visits, practical demos, and collaborations. The sales model broke during the pandemic because of the ban on in-person meetings.

Virtualization has been replacing in-person meetings and interactions. Scarce specialized skills are becoming more abundant. Physical locations and real estate have become less important. And travel and its related environmental toll are minimized.

Orthodoxy 4: Cash Is King

By dint of history, as the largest producer and consumer of hydrocarbons, the US economy provides the currency of the industry, and the US banking system and its capital market regulators play the de facto supervisory role over the commercial dealings of the global sector. While there are considerable benefits conferred by the stability of this state of affairs, there are some shortcomings.

The oil industry is impacted by the US government and its monetary policy, and the actions of the US Treasury can be at odds with the interests of the global oil industry. For example, the US independently manages money supply to suit its broader economy through mechanisms like adding to money supply (quantitative easing), which causes devaluation, controls inflationary risks, and helps manage banking. Those economies that wish to access global oil markets must maintain healthy foreign exchange balances.

Environmental exposure now limits the capital availability to the industry, inadvertently creating winners, not on environmental grounds but on capital access.

Digital tools offer a pathway to access capital outside of the US financial system. They create a mechanism to separate the ownership of industry assets from the operators, an opportunity for a new transactional currency to emerge, and the potential creation of entirely new asset classes.

Orthodoxy 5: Trust, But Verify

The oil and gas industry is marked by the absence of trust between supply chain participants. Oil in its crude form and refined petroleum products from both oil and gas are fungible, unmarked, and prone to theft. The high margins of the products combined with the huge volumes involved create plenty of commercial scope for fraud, kickbacks, and graft. Services take place at distributed and often unmanned facilities, creating the conditions for low quality and poor performance. Deep technical skills are often necessary to appraise the worth of work.

The scale and complexity of the industry leaves plenty of opportunity for honest errors to creep in. **Royalty** payments may be based on paper contracts signed long ago. Mechanical failures create disruptions in the normal flow of products. The sheer number of participants suggests that at least some portion of well-intentioned transactions end in error.

To maintain control over this potential commercial chaos, the industry has invested in the kinds of back-office capabilities that match the best in the world. The outlay for ERP solutions like SAP is in the billions for the whole of the oil and gas sector. Agents abound to facilitate transactions and trade. Some processes in many nations are even prescribed by law, requiring such outdated artifacts as manual custom stamps, physical signatures, multiple physical copies, and in-person settlement. Overall, this is a costly approach to business.

COMPLEX PROCUREMENT is not just a feature of large companies in this industry. In 2021, I published a few articles and podcasts for a small oil and gas software company. For my few hours of time, I was required to sign a non-disclosure agreement, a master services agreement, and a statement of work agreement, which were entirely disproportionate to the work being done and the procurement risk. Invoices were necessary to trigger payment on forty-five-day terms. Once their invoices reached seventy-five days overdue, they requested we settle via credit card, which triggered an unfavorable currency charge of 4.6 percent or $184 to the company. They then asked that the entire transaction be reversed and redone. It almost certainly cost more to process the invoice twice than it did to write off the currency adjustment.

In the same way that Uber reimagined the relationship between the passenger and the driver, digital solutions could remake the relationship structures in the supply chain to lower the cost of trust, enable gain sharing, reduce transactional friction, and allow a more equitable distribution of value.

Orthodoxy 6: The Consumer Is Anonymous

The relationship between energy producers with their end consumers is weak at best, and often downright hostile. Consider the reaction by the Texas energy companies towards their customers during the freak winter storm early in 2021.[3] With a simultaneous drop in energy supply with natural gas assets and wind farms freezing over, energy costs skyrocketed to well over $9,000 per megawatt hour. In typical utility fashion, they requested that the consumer manage their consumption (despite how absurd it was to have done so given the exceptional circumstances). Now they are expected to forgive energy bills as a result. None of this speaks to a particularly loyal relationship between provider and consumer.

There's little wonder why, however. Other than for their gasoline needs, energy customers don't make frequent choices about energy and rarely even think about their energy companies. Household energy budgets become fragmented across power, heat, and fuel. Similarly, businesses lack a single point of energy accountability. The chief procurement officer may negotiate pricing, but rare is the organization with a chief energy officer who worries about efficient consumption.

Commercial systems for supplying energy to the world cater to this construct and have not materially advanced in a generation. Many energy companies are quite distant from the actual consumer, and the needs of the consumer are not always visible or even understood. In many cases, the actual consumer is anonymous and, increasingly, a smart machine. Oil and gas companies are quick to point out that they don't even have customers; they have business partners.

A similar challenge faces automakers, who historically have placed the customer relationship in the hands of the car dealership. Tesla has disrupted this cozy structure by selling directly to customers and maintaining an ongoing relationship through continuous software upgrades.

The arrival of battery-powered transportation from the major auto brands suddenly enables a fresh competitor for the fuel relationship, namely, the local power utility who has supply contracts in place and a natural advantage in standing up all the charging stations.

The relationship that consumers have with their various energy suppliers, for the first time since market deregulation, is now in play. Increasingly that relationship will be machine to machine, another field of advantage for digital companies.

Orthodoxy 7: Energy Sources Are Irrelevant

Of all species, humans have a curious fascination with the unusual and exotic, and many among us are great collectors of things. We tell stories about where our photos were taken, about the artist we met who carved a mantle figure from a piece of bone, or about a colorful stone we found on a favorite beach.

This fascination with place of origin permeates our relationship with many things. We purchase organic foods in ever increasing volumes for the supposed health benefits. Apps provide us with a direct line to the merino sheep whose wool became the fibers in our sweater. Geographical indicators, for products such as Scotch whisky, Parma ham, Champagne, and Edam cheese, confer authenticity and guarantees of quality, have the backing of courts, and protect suppliers from counterfeit products.[4]

Energy systems have not been designed with consumer choice in mind. Energy value chains tend to be cost- and margin-based, and not consumer-based. Energy products are pushed down the chains, and consumers rarely have the ability to choose between energy sources at the margin. Choices in power in particular are characterized by long-term contracts and supply commitments.

Consumers are being conditioned to have and to exercise choice, in areas as varied as where they shop, the vacations they take, and the clothes they wear. This same expectation of choice around energy selection for price, quality, source, or carbon content is real and becoming codified in regulation.

Orthodoxy 8: Green Energy Is a Premium Product

An economist defines a luxury good as a good whose demand rises more than proportionately with rises in income—the richer you are the more you buy it. Being "green" absolutely meets this definition. To be truly green, we need to spend much more for the energy basics— heat, light, fuel for transportation—and a whole lot more to ensure that all consumables (food, clothing, everything) is also purely green. It's possible, but expensive. It is absurd to expect masses of consumers to pay more for a purely cleaner fuel while lower-cost alternatives exist.

SOME KINDS of green energy meet my more uneducated definition. To me, a luxury good is typified by its scarcity and exclusivity by which it can command a high price. It has a limited market of price insensitive buyers. It is made in small lot sizes, using specialty manufacturing techniques. It meets the highest standards of cleanliness, purity, simplicity, functionality. It inspires a degree of awe, perhaps from its beauty. It's viewed as highly desirable, but unaffordable to a significant portion of the population.

Luxury goods usually have readily available, perfectly good, and usually far cheaper alternatives. I could buy a Tesla, but instead I'm driving a much cheaper, used gas-powered Subaru.

Pure renewable green energy is a triple-A fuel—available, affordable, and abundant. Wind and sunshine are clean, relative to fossil fuels. They produce zero emissions. They are safe to handle (exposure to the sun and wind can be harmful, but usually not). They are imprecise energies unless first converted to electricity, after which, as power, they can be delivered exactly where you want at your desired intensity. Electricity is highly controllable, well understood, and reliable. It's fast (as in it's instantly on). It's highly desirable—as nations get richer, their populations are prepared to pay for cleaner energy to provide for cleaner air.

Added bonus—green energy reuses the legacy copper wiring, plugs, and appliances.

But there still are many barriers to wider adoption of green energy in advanced economies:

- a legacy building stock not designed for renewable energy capture;

- local building codes that permit hidden gas lines but prohibit "ugly" solar panels and wind turbines;

- financing models that foist the cost of retrofitting on the building owner;

- capacity constraints in building trades that inflate the price of an energy system overhaul;

- rental markets that block structural improvements or rent increases; and

- a grid system designed around a few large, centralized power generators, unable to cope with energy supply variability and widely distributed generation.

The price of renewable energy now rivals **gray energy** in all known markets. Solar and wind energy are beneficiaries of Moore's Law of exponential change. The more we manufacture renewable energy collectors, the cheaper they get, and the more efficient they become.

Digital innovations cannot overcome all of the limitations of green energy adoption, but they can make a big difference. Smarter, digitally enabled grid systems should allow variable energy supplies to play a role. Digital technologies allow for fractional ownership and financing of the capital required (solar capacity, storage, and transportation). Real estate developers could be required to incorporate community utility models. New houses can be built with solar panels and localized energy storage in mind.

Orthodoxy 9: Own the Assets

In 2015 I was meeting with an executive at a Queensland gas producer. He was griping that his team, yet again, submitted a purchase requisition for several new pickup trucks. Why, he asked, did they always want to *buy* the trucks, and why did they have to be *new*? At that time, the price of oil had fallen, and along with it the trading price of natural gas (which was tied to the price of LNG), and cash was scarce. He scribbled "not approved, lease used" on the requisition and sent it back.

This little exchange illustrates an important rule. The pattern in the industry is to buy, to buy new, and frequently to buy double in case the first purchase doesn't work out. This probably makes sense when the industry is in growth mode, assets are scarce, lead times are long, and capital is loose. It has even created a cottage industry of entrepreneurs trying to monetize the excess inventory of assets and spares that clog up warehouses and laydown yards.

Historically, owning the key assets of the industry was much less administratively burdensome than renting or sharing those assets. But many other industries have been experimenting with asset-light business models where the operator of the principal asset doesn't actually need to own the asset itself, and the assets move to another party's balance sheet. This is the case with aircraft, ships, taxis, and of course rental cars. The conditions are right, and the technology is now available, to refashion balance sheets in oil and gas.

New Rules for New Models

Entrepreneurs, both inside the incumbent players as well as leaders of startups, are challenging these orthodoxies and creating new business models. To my knowledge, none have yet reached the stage to be declared a unicorn, but watch this space. As the Uber example illustrates, there is not a one-to-one relationship between orthodoxies and successful business models. In fact, the more the rules are upended in the aggregate within a business model, the greater the success.

Data-Driven Businesses

Data business models are centered on the data of the industry, not the molecules; they are evolving dramatically and at pace.

INDUSTRIAL CUT-AND-PASTE

My first exposure to the problems and opportunities afforded by oil and gas data date back to the mid-1990s. I was carrying out workflow analysis for a seismic processing company that was facing some very irate customers. The company offered a service to convert 2D seismic data captured on large-reel magnetic tape to small IBM data cassettes. The reels were suffering from **stiction** and quickly deteriorating in the warehouse environment. The service was failing because the company insisted on carting off a large truckload of reels from the customer's warehouse to the processing center, assuring a large inventory of tapes for the conversion teams, but preventing customer access to their data. Inevitably, the customer would phone, urgently needing a tape that was hidden somewhere in the processing center.

The fix we evolved was multifaceted—we substituted taxis for the truck run, limiting the batch size of tapes to be transferred to what would fit in the trunk of a car, freed up space in the processing center, and reorganized the floor of tape conversion workstations to improve flow. The problems literally vanished overnight.

Though the absolute amount of data was quite small, the volumes of tapes involved beggared belief. There were hundreds of thousands of reels of magnetic tape data to be processed for this one client, for only one of their warehouses, representing perhaps a fraction of the global market.

There was no interpretation of the data, no analysis of the content. This was industrial-grade cut-and-paste work.

REAL-TIME DATA INTERPRETATION

Ten years on from my seismic experience, I worked with a major fracking outfit that, as a way to high-grade their quality of service, connected all their frac spreads to their in-house control room. Frack data flowed continuously and in real time to this facility, frequently via expensive satellite uplinks. The company's best engineers and fracking experts spent their days in the control room huddled around screens, guiding the fracks, tuning the horsepower, dealing with upsets, and generally keeping the assets as utilized as possible. They did nothing with the data once the frack was over. Despite their sophisticated customer-facing analytics, they relied on an enormous whiteboard with magnetic labels to identify where their own frac spreads were located. The cleaning staff could throw the business into complete disarray just by erasing the board.

HARDWARE, NOT DATA

Fifteen years later, I worked with a **downhole** tool company in Brisbane whose measurement devices provided highly reliable production measurements—water pressures, temperatures, radioactivity, flow rates, volumes, and many other data points. They sold the tools to their customers (in this case, gas well operators), who used the data to manage well performance, schedule services, and forecast production. The problem was that as the price of gas fell (gas pricing was determined with reference to the price of oil in Asia), the demand for new gas wells dried up, and with it, demand for their downhole tools.

I asked them at the time, in an emerging world of cloud computing and analytics, why they didn't capture the data themselves and interpret it as a service for their customers. They could have built up an enormous library of well operating conditions. Too hard, they claimed, even though their own product testing regime included the analytics capabilities and dashboard that could have served as the basis for a customer portal. They also viewed the data as belonging to the customer.

DATA, NOT HARDWARE

In late 2020, I ran my digital awareness training course for a Middle Eastern downhole tools company. Their tools and technologies were deployed in perhaps two thousand wells worldwide, mostly in highly productive offshore sites. Their kit produced a staggering amount of data, and they supplied their customers with the necessary analytic software and dashboards for its interpretation. Their analysis was first class, which explained their premium pricing and extraordinarily small market share. Data rights were again flagged as an area of uncertainty.

The prize was to figure out how to give their downhole sensor technology away to the three million available wells so they could capture as much data as possible, and to charge for an automated data interpretation service.

These days, hardware suppliers and field services companies are attentive to the potential value of the data resources produced by their operations. President and CEO Jim Rakievich, from McCoy Global, poignantly expresses the value of data thusly: "For ten years we sold expensive tools and equipment. I lose sleep now that I have a better understanding of the value of the ten years of data that we don't have."

Contract clauses are now framing the data as part of a service offering. Using AI, a fracking firm like Calfrac Well Systems, a drilling services company, gets smarter and smarter over time about execution because it has accumulated hundreds of drilling histories. GE does the same with its turbines.

Data Merchandizing

A second intriguing business model involves the **monetization** of data. Oil and gas has long sold data in the form of seismic surveys, but also in reports and datasets. Reports in paper format are secure products in that they're hard to copy, but raw data is another matter. Custody transfer, search and retrieval, and physical formats are clunky.

Enter the transformative potential of the cloud.

DATA, THE PRODUCT

Netflix and iTunes have shown the way for handling very large datasets and renting or selling them as a commercial business. The model does not precisely transfer to oil and gas, as the consuming market is much smaller and the customer wants to carry out their own analysis on the data. Unlike movies, which have just a few standard formats, data in oil and gas is wildly diverse. Nevertheless, the early contours of this model are taking shape. "You'll be surprised at what people will be willing to pay for data," says Jim Rakievich.

The data platform needs the capability to ingest the full range of data from the many sources that the industry uses, including SCADA historians, **well log**s, legacy software, subsurface data, Excel spreadsheets, ASCII files, PDFs, and drawings. The ingestion engine needs to identify and correct anomalies and build taxonomy. The platform will be continuously expanding and structured for ease of searching, analysis, and data manipulation. At scale, the repository will have enough data to enable value-added services, including machine learning, analytics, and AI. Third parties will be encouraged to produce their own apps for leveraging the data, using advanced technologies such as virtual reality, in much the same way that Apple's App Store unlocks innovation for its iPhone and iPad universe. Security is paramount for the industry, and data platforms will feature robust security and encryption services to prevent unauthorized access, use, and distribution of data. Game-like features, such as rankings for very useful datasets, apps, and algorithms, will round out these platforms.

Data platforms will need to incorporate the kinds of commercial elements that make the online marketplaces so appealing, including all sides of the market (suppliers of data, consumers of data, and services from apps). Careful tracking of ownership rights, over data and algorithms, as well as for the sale and transfer of rights, will be required. As data is consumed (either as a stream or as a download with restricted usage rights), the platform will need various mechanisms to settle the commercial terms.

Owners of datasets that could have value but lack the ambition or skills to monetize the data could use such a platform as a monetization mechanism. Customers of data (analytics engines, data displays, algorithm developers) could find such a platform an ideal place to develop solutions. Otherwise, they'll have a hard time accessing these rich data resources.

I suspect the first datasets in the repository will be comprehensive, desirable data with known value, such as seismic data. Oil and gas companies have plenty that they lack the capacity to analyze. They won't need to invest much time in clarifying the context of the data. Financiers understand the value of this kind of data and how to value the business. The base technologies (cloud storage, blockchain registers, security, streaming) are well understood and technically sound.

THE OPEN DATA MOVEMENT

Using Google Earth, pay a visit to Midland, Texas, the center of the prolific **Permian Basin**. Scroll around in the satellite view and visually digest the scale of the oil and gas production in this corner of the sector. All those tiny squares are **well pad**s. Zoom in for a close-up of the tanks that hold oil, water, and chemicals. Here and there you may note slightly larger facilities, the gas and oil plants, that aggregate production for a small area.

Visibly, this landscape is concerning because of the magnitude of the land disturbance from all the roads and well pads when far fewer were necessary. The invisible problem is the highly fragmented ownership of all these wells. In the US, landowners hold the rights to the resources under their feet (unlike Commonwealth countries, in which the resources are owned by the nation), and many farmers struck deals with the first oil and gas producer at the farmhouse door pitching to extract those resources. This is an inefficient model for basin development.

The solution is easy enough—buy out the competitors in an area, build fewer but slightly larger pads with multiple wells, and implement cost reduction and productivity improvement measures across the lot. The problem emerges when buyer and seller try to arrive at a value, and each are relying on dramatically different datasets trapped in application silos to determine that value.

Open data standards are how many industries solve this problem. Data standards allow for the rapid and trusted transfer of data between parties. Imagine a music world where there was no acceptable standard or language to write music. Musicians could not convene into orchestras. What if there was no MP3 file standard? Music sharing as we know it today on iTunes, YouTube, Amazon, or Spotify could not exist.

An open data movement called the Open Group OSDU Forum is a vendor neutral environment for the development of open data standards for oil and gas. The movement is based on the recognition that the lack of a common data platform architectural design is at the root of much industry inefficiency. The OSDU is developing an open platform for sharing industry data.

Vendors are likely feeling alarmed by this turn. It is a long-established vendor practice in oil and gas to create impenetrable walls around proprietary technology, block small technology companies from expanding, buy out entrepreneurial innovators, and extract rent through the customer relationship. But if the experience from other industries is any guide, open data is key to unlocking innovation and growth.

People-Lite Businesses

The dangerous nature of petroleum creates a permanent pressure to remove the human element from the work processes. Digital innovations in automation and autonomy are finally allowing the industry to tackle its people-centered orthodoxies and move towards the safest possible extraction and processing business model that features ever fewer people. And not just in the frontline, but also in engineering, design, trading, back office, and services.

AUTOMATION AND AUTONOMY

One of my must-have features in cars that I own is cruise control. I find it tiring to constantly press down on the accelerator and adjust speed on long highway drives. My concentration wavers, and I find myself chronically speeding. Cruise control is a nifty mechanical automation feature that removes the need for my constant supervision, but it has some limitations. It doesn't react to the presence of cars in front of me, and on downhill road sections, it inefficiently engages the engine and not the brakes to slow the vehicle.

Autonomation is a more intelligent form of automation. An autonomous machine carries out a mechanical job but is also able to rapidly detect, address, and correct for possible mistakes or variances from normal. This relieves the human from continuously having to judge if the machine is operating properly. Work speeds up dramatically. In factories that are highly autonomous, workers supervise multiple machines simultaneously.

Heavy industry in general, and manufacturing in particular, has been pursuing the automation-to-autonomation agenda for many years in response to competitive pressures, the quest for higher quality, and the need for improved safety outcomes, and in pursuit of lower environmental impacts. Jobs that embody the four Ds—dangerous, dirty, distant, and dull—are targets for autonomy. The imperative and opportunity to bring greater levels of autonomous operations to oil and gas have been given a further boost during the COVID pandemic. Oil and gas will follow the mining industry in adopting greater levels of autonomy, but even partially autonomous businesses achieve better economics, which allows them to grow faster than the market.

AUTOMATED DESIGN

Automation still has tremendous play in oil and gas because so much of the industry is still manual.

While I was working in Australia, I came across a company whose business was to translate engineering content into the instructions for factories to form and cut parts. For example, a tank design has a stairway attached to the side to give workers access to the top. At some stage, this design needs precise instructions on how to shape the handrail and where to drill holes to attach the handrail to the stairs. The tank design is from an engineering house in expensive Australia, but the instructions for the machine that cuts the steel and drills the holes is completed in a low-cost offshore design center.

Unsurprisingly, the offshore design center adopts automation tools to make their work faster and with fewer errors, and the Australian engineers push more bulk engineering work their way for such areas as piping and spooling. Meanwhile, engineering software itself becomes more capable and progressively automates the low-end bulk engineering work entirely.

As it turns out, much engineering work appears amenable to this kind of rules-based automation using digital tools. Such a shift has an outsized impact on an industry that is very much based on large numbers of engineers toiling away on dollar-per-hour contracts. The apprenticeship learning model in engineering (juniors supervised by seniors) fails when software does the work of the junior. The industrial logic of the low-cost offshore engineering center falters if software eats the offshore center.

Automation of work that was formerly high-end skilled labor plays out at all stages in the industry:

- engineering design and analysis
- geologic analysis
- visual data interpretation
- equipment supervision
- operations supervision
- scheduling people and equipment optimally
- optimization of operating equipment

Oil and gas services businesses whose work is repetitive, rules-based, and contracted in bulk should be wary.

ROBOTIC RESOURCE EXTRACTION

Developments in the surface mining industry provide a clear view to the possibilities of robotic resource extraction. Australia's mighty Pilbara mining region is home to two dozen major iron ore and manganese operations, and Australia extracts many other ores and minerals, and natural gas. These resource businesses are the global leaders in resource innovation, and their autonomy successes have been brought already to the oil sands in Alberta.

The benefits reported by autonomous operations include

- standardized operations reducing inconsistencies, errors, and delays;

- optimization of specific tasks, leading to cost savings and enhanced asset capacity;

- removal of human direct control of machinery, reducing errors, damage, accidents, and delays;

- better safety performance and improved resilience, minimizing the impacts of disruptive events;

- better quality data and analysis collected from autonomous machines leading to reductions in non-productive time; and

- efficiency gains leading to emissions reductions.

Canada's oil sands industry has dutifully followed the Australian miners and deployed these same autonomous solutions in the big open-cut **bitumen** mines. These mines are also in a fire-prone area (Canada's vast arboreal forests) and have been overly impacted by the pandemic, adding to the incentive to remove people from the business model. Unlike iron ore mines, some oil sands producers partially refine the ore locally. The oil sands mines have the scale, access to relevant suppliers, the cost incentive, the integration, and the talent incentive to create fully autonomous mines along with the Australians.

Equinor has pioneered the first autonomous offshore production platform, the Oseberg Vestflanken H, to tackle the Oseberg oil field in the North Sea.[5] The platform was built 20 percent faster than planned, smaller than its peers, and several tons lighter because it lacks any accoutrements for humans, including toilets.

Autonomous drilling is still a work in progress, plagued by incompatible rig subsystems, lack of open data standards, traditional contracting models, and few suppliers of robotic tools and subsystems, and will still be largely human-powered for the foreseeable future.

WHY REMOVE the human workforce from operations? Humans are expensive, unreliable, and buggy. They need climate controlled operating conditions (while oil and gas assets are often in extreme settings). Humans need to be protected from drops, trips, vapors, heat, explosions, sharps, and chemicals (standard operating features of energy infrastructure). Humans' ability to concentrate degrades over time (energy systems run round the clock). Humans take breaks (assets do, too, but infrequently). People trial dangerous shortcuts without thinking about the consequences (and bad things happen). People are prone to bouts of intoxication and revelry (fun at a party, but unwelcome in an industrial setting).

Midstream assets such as oil refineries, **gas plant**s, and pipeline operations are already highly automated, leaving little room for greater improvements from autonomy in operations, although late in 2020, Shell had acquired two four-legged robot dogs from Boston Dynamics for experiments.[6]

THE AUTOMATED BACK OFFICE

A highly automated back-office function in an upstream oil and gas company has a significantly lower cost structure than its peers, enabling it to grow aggressively through acquisition in the absence of access to traditional capital.

The earliest example of this model was NAL Resources, which adopted a handful of digital technologies to help it weather the 2014–2015 oil market collapse. Chief among these was RPA.[7] NAL discovered, much like other users of RPA, that the benefits were dramatic.

ROBOTIC SERVICES

The many services companies that work in the industry are at greater risk of the impacts from autonomation than their customers. Competitively, services are more easily transformed because their contracts turn over regularly, creating windows for alternatives. Services often have large numbers of employees whose work is amenable to autonomation.

The emerging business model for many services companies is to transform their operations to incorporate robots in the place of people. Drilling services are a specific example, but there are many more:

Earth works: The major equipment manufacturers, such as Caterpillar, Komatsu, Hitachi, and Volvo, have either delivered or are developing robotic and electric versions of excavators, loaders, haulers, drills, and dozers.

Logistics: Semi-trailer trucks, the work horses of the North American trucking industry, are trialing electric motors and adopting autonomous features. Light duty trucks will benefit from the same autonomous technology that has appeared on cars.

Underwater: Many jobs carried out on seabed installations (inspections, minor repairs, maintenance, cleaning and servicing) can be partially or wholly completed by autonomous vehicles and drones.

Inspections and surveillance: UAVs carry out overflights of facilities, pipeline right of ways, and development areas at a cost that is dramatically lower and considerably less risky than helicopters and light planes. Photographs from the next generation of satellites interpreted by machines introduces the prospect of near continuous facility monitoring.

Services companies that incorporate more robotic solutions into their business models are able to scale their businesses once they are freed from the tyranny of a human-centric model. Robots are a natural platform for collecting useful data that feeds deeper analytics about the business.

The robot-driven, autonomous service provider of the future will be able to

- deliver a higher-quality and more consistent service;
- execute work faster, depending on the service;
- boost the productivity of service assets;
- create a more attractive work option for the next generation of talent;
- lower the emissions footprint from more efficient operations;
- improve resilience to pandemic events;
- lower the cost of service with fewer staff on-site;
- produce new revenue streams from software and usage; and
- unlock creative financing models based on robotic assets.

ROBOTIC OPERATIONS

The bigger prize, yet to be fully realized, is turning onshore production assets into robotic production units. Production assets are strong candidates for autonomation. These assets are often well beyond the reach of modern telecom networks and cannot be run using cloud solutions. They are almost always wired up with SCADA systems because the case to supervise these assets is already sound. They often suffer from long delays in management intervention because their processes are so manual. Their costs are high because of the need to send people to them on a regular (daily) basis. The few operators who have introduced edge computing to production assets have seen solid performance gains—more production from existing assets, fewer personnel per operating well, greater leverage for staff, enhanced staff decision-making, and an objective view of the asset that is unfiltered by spreadsheets and human bias.

Any greenfield upstream oil and gas business is in a position to design itself as a true autonomous producer and to lower the cost of production through digital.

New Financial Structures

There is no good reason that oil, a global product used by every country and widely produced by dozens of countries and hundreds of companies, should be tied to a single sovereign currency.

And so it wasn't that long ago that the Venezuelan government attempted to launch their own cryptocurrency, called the petro, backed by their oil reserves in 2017. Venezuela has historically been a serious oil player.[8] They're a founding member of OPEC, and the country is blessed with reserves of some 300 billion barrels of oil it could produce, estimated to be the largest reserves in the world. The Orinoco Belt has additional reserves of very heavy oil, amounting to 1.5 trillion barrels.[9] About 95 percent of their export earnings come from the sale of oil and gas, and the sector is a full 25 percent of their GDP.

Venezuela has fallen on hard times, and its **crypto** idea has failed, but the industrial logic behind it has endured. Many oil-producing economies, such as Venezuela's, are tightly interwoven with the American dollar. Oil-producing nations need USD to pay down national debt borrowed on US capital markets. Their principal (and often only) export is priced in USD, and production and operations can depend on international expertise and equipment valued in USD. Sanctions can immediately freeze a country from accessing the global banking system, which is dominated by the US.

Creating a **token** that represents a specific barrel of crude oil or a cube of gas has much appeal if you're attempting to disentangle your commerce from the US financial system. Cryptocurrencies look like they might be a solution to this dependency on USD, but uncertainties remain:

- Not all oil is equal in value. Light **sweet** oil comprises the benchmark oil indexes, but heavy **sour** oil trades at a discount because it's more costly to refine. An asset-backed token needs to reflect the fact that the underlying oil or gas asset is not purely fungible.

- With the value of the underlying asset varying relative to the quality of the asset, the value of the token will likely need to vary.

- A token representing an asset may well have its own separate value, distinct from the underlying asset. Logically, the price of the token cannot fall below the value of the asset, but commodity markets have experienced negative oil prices when storage runs out. Rules may be needed to manage volatility.

- Traders in an oil-backed cryptocurrency need confidence that they can convert their crypto holdings into a real asset, such as crude oil. A physical backstop becomes important.

- The rules and mechanisms for converting a token representing a physical barrel need to be set out. For example, once the barrel of crude is converted into petroleum, the token must be destroyed or archived.

- A token representation of oil becomes a player in futures markets. Most buying and selling of oil is conducted using futures contracts where traders structure deals to maximize profits and minimize losses by transacting oil forward. More rules become necessary.

All of the interest in cryptocurrencies implies that, sooner or later, another, more successful attempt to figure out how to transact for oil and gas using tokens will take place. Meanwhile, simpler ideas that allow for trials and experimentation will emerge.

For example, one company uses tokens in a dual market model on abandoned wells. Thousands of abandoned wells are shut in because conventional technology to maintain their productivity is too costly for the one to three barrels per day of production they deliver, and capital is not attracted to this opportunity. Beyond Oil targets these wells with its low-cost pneumatic lift system that uses solar power to compress air for the pneumatic power.[10] Investors hold tokens representing their investment, which pays for the pneumatic system. Each barrel produced is captured as a token; when the barrel is sold, the token is destroyed and the value is returned to the investor.

This model accesses capital, solves for an environmental liability, puts a resource to work, and leverages clean technology. It also creates an opportunity for the industry to learn how best to apply tokenization technology to the physical industry.

Supply Chain Tracing

Sometimes it takes a shock to a system to accelerate a change already underway. A series of shocks confronted the trust problem in the industry and brought the concept of track and trace to the infrastructure sector. New solutions are helping create cradle-to-grave visibility to the assets and the products of the industry.

WHEN DISASTERS HAPPEN

A succession of North American pipeline failures triggered a regulatory response of significance for the industry. In July 2010, a forty-foot segment of pipe in Enbridge's Line 6B ruptured, spilling diluted bitumen into Talmadge Creek and on into the Kalamazoo River in Michigan.[11] Two months later, a gas pipeline operated by Pacific Gas and Electric ruptured in a massive fireball in San Bruno, near San Francisco.[12] This conflagration killed eight, and leveled thirty-five homes. In April 2011, Plains Midstream experienced a pipeline rupture on its forty-four-year-old Rainbow system near Little Buffalo, Alberta, causing the largest spill in Alberta's history.[13]

Regulators on both sides of the Canada-US border were stunned to learn that the pipeline companies were unable to quickly respond to basic questions about their pipelines: 35 percent of companies questioned could not say definitively what pipes were installed where, which mill manufactured the pipes, who welded them together, the conditions at installation, or the history of inspections. The conditions of the pipes were unknown.

For an industry that prides itself on operations excellence, and whose brand depends on reliability, this was a huge wake-up call.

The regulators quickly advanced new rules. The Pipeline and Hazardous Materials Safety Administration (PHMSA) in the US and the Canadian Energy Regulator (CER) published their new rule books. In the US, PHMSA rule 192.67 requires pipeline material property records to be complete and accurate.[14] In Canada, CSA Z662 is composed of over five hundred pages of prescriptive and performance-based requirements and covers the technical aspects of design, construction, operation, maintenance, deactivation, and abandonment of oil and gas industry pipeline systems.[15]

The biggest change to the code is that material records (about pipe segments, welds, and other equipment) must be traceable, verifiable, complete, and accurate. Put another way, pipeline owners need to be able to show the full lifecycle of an individual segment of pipe, from source to location, to condition, to disposal.

Traceability is not the same as tracking. Tracking means knowing where something is. Tracing means knowing who or what you are, where you have been, what you have been doing, and where you are presently. Tracing is harder and needs a lot more data to do but, done well, yields a lot more useful insight.

Particularly for the pipeline industry, tracing is a profoundly difficult problem to solve for a number of reasons.

Vast scale: Canada alone has 840,000 kilometers (nearly 522,000 miles) of pipeline.[16] Pipe is sold in various lengths, but a common length is forty feet, or 12.2 meters. There are eighty-two segments of forty-foot pipe per kilometer, yielding some 68 million segments overall. There are at least 68 million welds holding these pipes together. The US has at least ten times more pipeline than Canada.[17]

Diverse locations: Most land pipelines are buried a few feet underground, so as to keep them from interfering with animal migrations or inadvertent damage from trees falling and fires. Pipelines cross the full gamut of terrain: over low valleys and high mountain passes, under rivers and lakes, through built-up areas, and across seabeds. Inspecting them poses its own challenge. Most pipes were installed before the invention of modern GPS systems and are tough to precisely locate. Older maps are paper and may not be up to date. Undersea pipelines are exactly that—on sea floors and hard to access.

Long life: Pipelines can work reliably for decades. Over time, segments of pipelines will be replaced, expanded, shut down, and converted from transporting liquids to gas and back.

Supply chain complexity: Hundreds of suppliers of **tubular** products from all around the world vie for business. Steel is the main material, but there are no universally accepted or fixed ways to characterize pipe products. Hundreds more companies move, lay, and weld pipe.

Product complexity: Not all pipes are the same. Pipe is sold in many different diameters and wall thickness, with different metals and different coatings.

Data sources: The data about new **linear infrastructure** originates in many different systems, with many different layouts, taxonomies, and terminologies. Pipeline purchase contracts may be silent on when the data about an order should be delivered, the format the data should take, and what the data should include.

TRACING ASSETS

Tracing of physical assets is simpler than the need to trace molecules of petroleum or carbon emissions. Nevertheless, a midstream company has estimated that as much as six hundred hours of people time is spent to prepare and manage the data associated with every million dollars of their capital spend, according to a confidential midstream source. Those are not cheap hours—an external consulting engineering rate of $150 per hour is not unreasonable, which, over six hundred hours, comes to $90,000, or 9 percent of the capital spend. It also slows down projects, causes delays, and inflates project costs. Better-quality data about pipeline infrastructure will trim that cost budget.

Operations can struggle to take accountability for a new asset that has uncertainties in the data. The longer the handover, the more delayed the revenue. The better the data, and the more confidence operations has in the data, the faster the inspections, testing, commissioning, and approvals.

Poor-quality data about pipelines now detracts from value. Companies that buy pipelines will discount the value of the pipeline asset by 30 to 40 percent if it is not accompanied by rock-solid data about the installed pipe, components, and welds. This implies that companies with pipeline assets on their books that are not backed up with quality data are potentially overstating the value of those assets.

Finally, in litigation, it is not necessary to prove to a jury that a specific piece of data is inaccurate or incomplete. Merely demonstrating that the system for managing data quality is error-prone can discredit all the data.

There are a few ways to solve this tracing problem. One is to create a hub that resolves the many-to-many problem (many suppliers, many customers, many products, but no data standard). These kinds of hubs appear throughout oil and gas, solving problems like disposing of surplus assets, enabling product sales, or connecting people with work. Newer hubs focused on data often provide the following capabilities. They can

- ingest data from a source system, supplier, file, or repository;
- transform the data to a standard;
- provide for data clean-up, rationalization, and correction;
- enable customer access to the data in the standard;
- provide value-added services on the data;
- enable access from any device, anytime, anywhere; and
- provide security, access, and controls over the data.

Traceability is coming to many different kinds of assets, including tanks, terminals, and turbines, particularly as the world builds out trillions of dollars in new infrastructure.

PRODUCT TRACING

A particular asset that could benefit from traceability is the actual energy product itself. The industry does a pretty good job of tracking hydrocarbons, but tracing energy is another matter.

It's striking that we have not historically been as concerned about the traceability of our energy as we have been with whisky. In part, this is due to whisky being more fun, but is also due to electricity being a local good, produced by a local utility, perhaps owned by the community, and consumed close to supply. On the other hand, petroleum producers have claimed, with some justification, that tracing the source and use of liquids and gases is not practical, because these commodities are fungible and frequently blended.

Times are changing, and companies cannot properly respond to regulatory pressures without understanding where and how their operations and products contribute to climate degradation.

Markets are responding to this general trend to understand energy provenance by applying sourcing concepts to energy products.

Airlines offer consumers a fuel surcharge to purchase carbon offsets. Power companies sell green energy options such as green credits. Petroleum companies offer a green fuel option at the pump.[18] Chemical companies now track the use of their products throughout their supply chains in response to global brand pressures.[19]

Carbon targets compel companies to be much more interested in the source of all of their carbon emissions, the provenance of their hydrogen (was it produced using fossil fuels or solar power?), the origins of their biodiesel, and the carbon content in their products from all of their suppliers.

Energy systems have historically been a series of separate and independent value chains. These designs date back to the dawn of the modern energy age and are generally built to solve for monopoly effects of large distributed infrastructure and for the capital needs of the asset owner and investor. They are focused on managing the engineering problem of balancing physical supply networks, not on participating in a world of traceable energy.

Fortunately, the same digital innovations that have transformed many other markets (financial services, telecoms, entertainment) are about to have the same positive impacts on energy. These building blocks—the IoT, cloud computing, blockchain—allow for the tracing of energy products completely throughout their independent and increasingly interconnected value chains.

The essential ingredient to transforming energy is data. Historically, the energy industry has relied on its instrumentation and controlling systems (SCADA) to produce the data needed to manage energy supply and demand. The IIoT will provide for vast new data sources about energy, to be stored and processed in the cloud, and immutably and confidentially recorded on blockchain structures for consumers. Emerging AI engines will be able to process that data to make meaningful consumer decisions.

Many organizations are starting to exploit these new capabilities to reinvent energy supply chains. In addition to VAKT, Mavennet, a Toronto-based blockchain company, is supporting the Department of Homeland Security in tracing petroleum movements across borders.[20]

The Customer Relationship

Hand in hand with the requirement to trace energy through the supply chain to account for its provenance is the potential to capture the relationship with the energy customer. Tomas Malango envisages the possibility of a new and different relationship: "I want to create a fan instead of just a client."

Energy consumers are demanding, and regulators are insisting that producers supply products and services that meet social standards for emissions, pollution, sustainability, and safety—including energy. Once the market provides clear choice, consumers will select cleaner and more sustainable energy, sending a clear signal back to the market and accelerating energy transition. Fraudulent, misleading, or undesirable energy products from nations ambivalent to their social and environmental obligations will be detected, and brands will be able to protect their reputations and social mission by avoiding such damaging products.

The business opportunity is to radically reimagine the consumer relationship with energy and capture the evolving customer relationship. Many factors are driving the relationship, including new consumer types, changing attitudes, new energy products, and available and inexpensive technologies.

For example, imagine the household of the future, with a single smart energy hub at its center optimizing heat usage, turning unused lights off, purchasing desirable energy at the margin based on the customer's criteria, running energy-intense appliances in off-peak hours, and selling back any surplus locally generated power to the market. This model becomes real very quickly once a battery car takes over the garage and solar panels appear on the roof.

It's not obvious that the local power utility will own the future. Oil and gas companies can be the spoilers in this looming winner-take-all energy relationship overhaul.

ENERGY TRACING

It is energy tracing that fully opens up the customer relationship. By adopting tracing functionality, the energy company of the future has access to the essential ingredient for reimagining the customer relationship, namely, data.

There is a profoundly better world at hand, enabled by digital technologies, where consumers have confidence in where their energy comes from. End-to-end energy transparency unlocks new markets and totally new models. With fully exercised and meaningful consumer choice comes deep consumer insight, a boon to energy companies whose relationship with their customers is a 1990s CRM.

THE REIMAGINED ENERGY COMPANY

Along with a different customer relationship, energy companies will also be able to reimagine the pillars of the energy business—O&M costs, financial accounting for assets, depreciation, and tolling models.

Capital allocation should improve as capital flows transparently to satisfy clear consumer choice. Operating costs should improve as asset maintenance regimes better reflect consumer needs for specific energy type availability. Gold-plating should diminish.

Finally, capital markets will gain. Market participants will redirect financial investments towards customer-responsive energy asset investments, which could unlock a new asset class of energy products. Capital markets may be able to more correctly value energy businesses based more on their revenues and margins, and not on their original cost or depreciation. Last, new business models, with the transformation impact of ride-sharing but on a multitrillion-dollar scale, can be realized; think guilt-free energy, charity energy, global energy surcharges, and fully democratized energy.

Reshaping the Balance Sheet

Balance sheets in asset-heavy industries have a lot of capital that is unavoidably tied up in hard assets, such as land, infrastructure, buildings, and facilities. Some of this capital can now be converted from a capital charge to an operating expense, which alters the capital model for the industry. Another emerging approach includes restructuring the business model to create more shared assets that otherwise would not be shared.

CAPEX CONVERTS TO OPEX

Converting **CAPEX** to **OPEX** brings a number of advantages. Simply reducing the capital at work helps raise returns, and freed up capital is often put to higher use. In oil and gas, physical assets tend to run to their volume, throughput, and energy limits, which means improvements on their returns are often tied to improvements in commodity pricing, and not the underlying design. Operating costs in comparison are much more impacted by competitive forces.

The returns on capital employed in parts of the industry, and shareholder returns, have lagged behind the overall market for some years now, and the industry is no longer that attractive for investors. Capital shortage is a serious cramp on growth.

The industry has long practiced this conversion tactic. Years ago, oil companies owned their ships, drilling rigs, trucks, and tanks, but these have since been flipped into either rentals or service contracts with services companies. Rental assets suffer from their own shortcomings, including a higher than necessary standby cost (when the asset is available, but not used), opaque rental agreements that are really a form of off-balance-sheet financing, and restrictive contracts. An unspeakable number of rental assets (power generators, tools, skids) simply walk off job sites every year.

The new model is to convert a capital cost into a subscription-based operating cost that is tied to usage or cycles. The consumer world is full of such examples, including streaming services, translation software, hosting charges, and all manner of cloud services.

In the industrial setting, examples include analytics services that exist in the cloud, robots that run services on demand, or functions that play out on a per unit basis, with modern logging tools like blockchain recording that usage immutably. Cloud-based services do not "walk away" as rental assets can.

Physical assets are still susceptible to pilferage, but imagine those dumb metal assets newly enabled with a cloud-based software solution that renders the physical asset inoperable without access to the cloud software. The incentive for the dishonest to abscond with the goods vanishes. Even relatively low-cost items like smart tools (hand tools with embedded digital smarts) can be beaconized, like using Apple's AirTag, so that the tool eventually shows up on the grid.

As for the company that provides these subscription assets, their financials show much less volatility and a steadier monthly cash flow, which improves their own access to capital and yields a better valuation. New financing models become possible, and along with money come new financial assets. Not to mention that a subscription asset also spins off data as a side benefit, and that data asset likely has value too.

The typical objection to overcome is that any outsourced asset, including the services of a subscription, or a rental asset, includes an embedded cost of capital charge, and, invariably, the cost of capital in oil and gas is less. My counter view is that while the cost of capital is probably less, oil and gas will not invest in that capital to keep it competitive, and will eventually accrue a technology debt to repay.

SHARED ASSETS

Digital innovations are permitting embedded physical assets in the industry to be split out and recast into shared assets. An example is the SCADA control room.

Regulations, particularly for pipelines, specify that the operator must be in full control of their asset at all times, which creates the business case for a central control room. The control room is an expensive asset in its own right. Like the pipeline, it operates 24/7, needs to be in regulatory compliance, and must function reliably under all conditions. Control rooms need special conditioning, such as redundant high bandwidth cabling and power. And, as the Colonial Pipeline cyberattack demonstrated, they must be physically and digitally secure.[21] Smaller pipelines lack the scale to offset the constantly rising costs of such facilities.

As SCADA companies move their platforms to work using cloud-based infrastructure, the logic of a shared control room becomes compelling.[22] A third-party service company can now operate multiple small pipeline assets as a service and capture scale economies by aggregating the pipeline networks under one facility. With scale comes significant operating benefits:

- Larger facilities are more appealing to talent with the prospects of career growth and exposure to advanced technology.

- Cyber costs can be spread out across more assets, and cyber solutions can be more quickly deployed to improve resilience.

- More assets under supervision capture more data, which creates the opportunity to apply machine tools to help with optimization.

- Upgrades are more readily deployed on cloud technologies.

- Shared facilities are more resilient to pandemic disruption since operators can be more easily redeployed if needed.

Control rooms are likely to become more virtual over time, larger and more comprehensive, connecting sensors and devices in new ways, and enabling machine tools to wring value from optimization. It will take a while, but the pathway looks clear.

Capital Execution

It is in **capital execution** where many orthodoxies of the industry could be overturned with some speed. First, the imperative to accelerate capital execution is pressing. The large engineering works that take many years to deliver and are typical of energy projects are now at odds with climate targets that stipulate dramatic improvements in just ten years. Second, the oil and gas industry is highly dependent on the construction industry since oil and gas needs to spend almost a trillion dollars every year to maintain and increase production to accommodate growth in demand. Even if demand declines, oil and gas will need new approaches to help.

CONSTRUCTION TRANSFORMATION

The front end of capital projects, namely, engineering, has always been an aggressive adopter of new ways of working—client pressures see to that. But the construction industry, even bigger than oil and gas with a global value of $12 trillion, has lagged behind almost all other sectors for many reasons:[23]

- The industry is labor-intense, often unionized, and generally low-skilled. It frequently makes more economic sense to add labor rather than make labor more efficient.

- The industry is highly fragmented, with hundreds of thousands of market participants. Deploying common digital solutions is very challenging.

- Construction is highly localized, with specific rules and regulations set down by local planning boards. Scale economies are hard to achieve in this setting.

- Participants in the sector have little incentive to adopt innovations given the unequal distribution of value from those innovations.

- The sector does not have a history of innovation, and this muscle is rather undeveloped. In 2010, the total amount of venture capital that entered the market for startups focused on the construction industry was just $8 million.

By 2019, venture funds were plowing $1.8 billion annually into various construction industry startups, and dedicated investment funds were finally taking aim at transforming the industry. Some of the biggest names in venture (SoftBank, for example) were targeting the sector. Digital startups in construction have achieved unicorn status (valuations of a billion dollars or more), including three-year-old Katerra, which declared bankruptcy in 2021.[24]

China has been forced to rethink construction because of its need to accelerate its industrialization to become rich as a nation before it grows old. YouTube features several videos of office tower builds, bridge construction, laying of high-speed rail, and even the assembly of COVID field hospitals in China at breathtaking speeds. Under the Belt and Road Initiative of rail, roads, and ports, China has built 15,500 miles (25,000 kilometers) of high-speed rail lines in the past decade.[25]

Meanwhile, the new HS2 project in the UK, a high-speed rail line connecting London to Birmingham and eventually Manchester and Leeds, has ballooned in cost from $56 billion to $143 billion (£56 billion to £106 billion).[26] Chinese companies have hinted they should be able to build out HS2 at much less cost, and I don't doubt it. While China's infrastructure successes have much to do with its authoritarianism, and the UK's issues often stem from legal, political, and social barriers, China clearly has an edge when it comes to construction sophistication from a digital perspective. And they aren't alone.

TOKYO HOSTED the pandemic-delayed 2020 Olympic Games in 2021, and as is typical of the Olympic movement, the city needed to build eight new facilities for the Games, set up ten temporary sites, and reuse or retrofit twenty-five others. But consider that the average age of a construction worker in Japan is sixty. One of the ways that Japan has managed to hit its impressive construction target is to use digital to boost productivity and lower the cost of construction.[27]

The IEA estimates that buildings account for a third of global energy demand and 55 percent of global electricity demand, and are a major source of GHGs.[28] Tackling climate issues means that buildings and their construction must become greener. New techniques include **unitization, modularization, stack construction**, wooden skyscrapers, and factory assembly.

THE VIRTUALIZATION OF ENGINEERING SERVICES

EPC (engineering procurement construction) firms had a latent virtual business model lurking in their midst, but it took the forced adoption of social distancing practices to prove out the merits of a virtual EPC business model, namely,

- rich access to resident specialist expertise almost anytime from anywhere in the world;

- better allocation of engineering work to locations best situated economically;

- deeper collaboration with peers, specialists, and clients;

- faster execution of work;

- swifter reviews, inspections, and approvals, conducted virtually;

- better access to libraries of shareable project data;

- greater levels of innovation;

- swifter adoption of digital services that roll out virtually;

- better load-balancing across offices;

- fewer large offices and a resultant lower-cost profile;

- reduction in costs and lost time associated with travel, in addition to lower carbon emissions; and

- greater pricing flexibility due to access to a more diverse and better distributed workforce.

The big change is the resourcing model, which shifted from resourcing by office to resourcing by capability. The era of a fully virtual global engineering and procurement company is here, and it will have significant structural cost and productivity advantages over the former model.

ENGINEERING THE DIGITAL TWIN

Underpinning the virtualization of the engineering business is the construction of a purely virtual version of a physical capital project or existing brownfield asset. Engineering firms have been carrying out digital designs for many years and slowly adding incrementally to this capability.

Once the digital version of an asset is released from its captivity within the walls of an engineering company to the cloud, this **digital**

twin can serve the needs of the entire supply chain of companies involved in its creation. Now all construction participants—designers, owners, builders, crafts, suppliers, and inspectors—work off a single set of data. The digital twin is used to model out constructability, fit components at site foundations, and mate factory assemblies across multiple factories. "As-built" data captured in the digital twin eases the hand-over process when the constructed facility is turned over to operations.

Much more work remains to be done, as shared by Azad Hessamodini: "Digital twin reminds me of an Indian parable of the blind men and an elephant; no one knows what a digital twin is, yet everyone has an opinion of what it is based on their limited and often narrow individual perspective."

In its growth pathway, the digital twin stretches beyond the narrow digital boundaries of a specific asset or a business throughout the industry value chain. This vision is within reach for a national oil company whose operations extend fully through the chain.

Meanwhile, there are still significant gaps in the supply chain, particularly between the EPC firm and its fabricators. Moving the fabricators fully onto the same digital platform means that designers can help avoid the staggering level of rework that plagues the industry.

Taking the concept of a digital twin fully through the supply chain brings these benefits:

- access to specialist expertise from any company, anytime from anywhere in the world;

- full allocation of engineering work to the best equipped supplier in the supply chain;

- collaboration and fast work execution across the entire supply chain;

- early identification of errors, emerging problems, bottlenecks, and constraints;

- reduction in disputes, rework, and delivery risks, leading to lower risk premiums;

- the ability to optimize the eventual operating asset as well as the construction schedule;

- greater levels of compliance with more complete data about the asset; and

- deeper analytic capabilities because of the availability of more information.

CONSTRUCTION FUTURES

The construction industry has ample opportunity to reinvigorate its business model with new technologies such as:

- smartphones with cameras and apps to capture safety incident data;

- robots that serve as worker companions and job assistants;

- robotic excavation equipment to improve earthworks productivity;

- augmented and virtual reality to provide for design walkthroughs, workforce training, and testing of emergency response;

- drones for aerial surveillance of construction sites;

- visual data interpretation to detect the arrival of time-sensitive deliveries;

- sensor kits to capture heat, humidity, and light conditions;

- digital marketplaces to source equipment, crews, and spare parts;

- digital tags and codes to track goods completely through the supply chain for warranty and fraud protection; and

- **3D printing** for structures, specialized parts, and site tools.

The oil and gas industry stands to benefit as these innovations make their way into the broader construction trade. What isn't yet clear is whether a new business model awaits these efforts.

KEY TAKEAWAYS

The business models in the oil and gas industry have been very stable for many years. Here are a few key takeaways from this review of the industry's orthodoxies and business model changes:

1   Many inherent business rules about oil and gas, from the value of data to the role of cash, are facing disruptive change enabled by digital technologies.

2   The relationship between the energy provider and the energy consumer is being reinvented on multiple dimensions, creating market entry opportunities for new energy competitors and services.

3   Alternative sources of capital are emerging that allow fossil fuel producers to sidestep traditional capital markets, raise funds, and reshape balance sheets.

4   As happened in other industries, new business models that appear in oil and gas are initially small and inconsequential. Capital suppliers are ready to invest.

5   Most of the new business models are within the supply chain of goods and services, and not within oil and gas producers, refiners, and distributors directly, highlighting an overlooked opportunity.

Notes

1   "Market Report—Avison Young Calgary," accessed June 29, 2021, avisonyoung.ca/web/calgary/market-report/-/article/2021/05/10/calgary-office-market-report-q1-202-1.

2   Elliot Smith and Ryan Browne, "How Traders Are Adapting to Life off the Trading Floor in a Global Pandemic," CNBC, May 1, 2020, cnbc.com/2020/05/01/coronavirus-traders-adapt-to-life-off-the-trading-floor.html.

3   Matthew Farmer, "Third Energy Firm Declares Bankruptcy in Texas Snow Storm Fallout," Power Technology, March 16, 2021, power-technology.com/news/industry-news/texas-snow-storm-bankrupt-fallout-energy-prices-ercot.

4   "Quality Schemes Explained," European Commission, 2021, ec.europa.eu/info/food-farming-fisheries/food-safety-and-quality/certification/quality-labels/quality-schemes-explained_en.

5   "Inside the First Fully Automated Offshore Platform," *Offshore Technology*, February 11, 2019, offshore-technology.com/features/inside-the-first-fully-automated-offshore-platform/.

6   Allison Bench, "Robot Dogs the Newest 'Employees' at Alberta Shell Refinery" *Global News*, March 7, 2021, globalnews.ca/news/7682274/robot-dogs-shell-scotford.

7   "An Interview with Cory Bergh and Michele Taylor," interview by Geoffrey Cann, Transcript, October 21, 2019, geoffreycann.com/interview-cory-bergh-michele-taylor.

8   Steven Melendez, "Russia, Venezuela, China Have Explored Using Blockchain to Evade Sanctions: Report," *Fast Company*, July 11, 2019, fastcompany.com/90375316/russia-venezuela-and-china-have-explored-using-blockchain-to-evade-sanctions-report.

9   Pierre-René Bauquis, "What Future for Extra Heavy Oil and Bitumen: The Orinoco Case," World Energy Council, 2006, web.archive.org/web/20070402100135/http://www.worldenergy.org/wec-geis/publications/default/tech_papers/17th_congress/3_1_04.asp.

10 Claudia Cattaneo, "'As Precious as the Resources:' Data Science Is Oil Industry's Next Big Thing," *Financial Post*, November 24, 2017, business.financialpost.com/commodities/energy/as-precious-as-the-resources-data-science-is-oil-industrys-next-big-thing.

11 "Enbridge Incorporated: Hazardous Liquid Pipeline Rupture and Release, Marshall, Michigan, July 25, 2010," Accident Report, National Transportation Safety Board, July 10, 2012, ntsb.gov/investigations/AccidentReports/Reports/PAR1201.pdf.

12 "NTSB Releases First of Six Factual Reports on San Bruno Pipeline Rupture Investigation," National Transportation Safety Board, January 21, 2011, web.

archive.org/web/20170519045344/https://www.ntsb.gov/news/press-releases/
pages/ntsb_releases_first_of_six_factual_reports_on_san_bruno_pipeline_rupture_
investigation.aspx.

13 Trevor Gemmell, Davis Sheremata, and Kerry Williamson, "Plains Midstream
Canada ULC, NPS 20 Rainbow Pipeline Failure April 28, 2011" (ERCB, February
28, 2013).

14 DeWitt Burdeaux, "PHMSA Safety of Gas Gathering and Transmission Rule,"
TRC PowerPoint presentation, accessed June 29, 2021, psc.nd.gov/jurisdiction/
pipelines/docs/2019%20Pipeline%20Safety%20Seminar/Gas-NPRM.pdf.

15 "CAN/CSA-Z662-99: Oil and Gas Pipeline Systems," Standards Council of Canada,
April 30, 1999, scc.ca/en/standardsdb/standards/7148.

16 "Key Facts on Canada's Pipelines," Minister of Natural Resources Canada, 2016,
nrcan.gc.ca/sites/www.nrcan.gc.ca/files/pipeline-facts/Key%20Facts%20on%20
Canada's%20pipelines_8_5x14-access_e.pdf.

17 Jacquelyn Pless, "Breaking It Down: Understanding the Terminology," National
Conference of State Legislatures, March 2011, ncsl.org/research/energy/state-gas-
pipelines.aspx.

18 "Biofuel Facts for the Road: The Energy Department and Your Gasoline Pump,"
Office of Energy Efficiency & Renewable Energy, November 24, 2015, energy.gov/
eere/articles/biofuel-facts-road-energy-department-and-your-gasoline-pump.

19 Anne Ter Braak, "Stahl Selects Finboot's Blockchain Solution MARCO," Stahl,
2020, stahl.com/news/finboot-sustainability-credentials-solution.

20 Press release, "DHS S&T Silicon Valley Innovation Program
Makes New Phase 1 Awards to a Global Cohort of Five Blockchain
Companies," Department of Homeland Security, October
9, 2020, dhs.gov/science-and-technology/news/2020/10/09/
news-release-dhs-st-svip-makes-new-phase-1-awards-five-blockchain-companies.

21 Raphael Satter, "Colonial Pipeline: What We Know and What We Don't about
the Cyberattack," Reuters, May 10, 2021, globalnews.ca/news/7848118/
colonial-pipeline-cyberattack-what-we-know.

22 "Iron-IQ's SCADA Revolution Press Release," Iron-IQ, March 5, 2021, iron-iq.
com/company-news/scada-press-release.

23 "Global Construction Industry Report 2021: $10.5 Trillion Growth
Opportunities by 2023—ResearchAndMarkets.com," Businesswire,
January 11, 2021, businesswire.com/news/home/20210111005587/
en/Global-Construction-Industry-Report-2021-10.5-Trillion-Growth-
Opportunities-by-2023; "Emerging Trends: Venture Capital Investment
in Construction Tech," CRETech, November 7, 2019, cretech.com/other/
emerging-trends-venture-capital-investment-in-construction-tech-2.

24 Chris Bryant, "How SoftBank's American House of Cards Collapsed,"
Bloomberg, June 8, 2021, bloomberg.com/opinion/articles/2021-06-08/
katerra-bankruptcy-how-softbank-s-american-house-of-cards-collapsed.

25   "China to Develop Driverless High-Speed Train," *Railway Pro Communication Platform* (blog), March 2, 2018, railwaypro.com/wp/china-develop-driverless-high-speed-train.

26   "HS2: When Will the Line Open and How Much Will It Cost?" BBC News, February 11, 2020, bbc.com/news/uk-16473296.

27   "Japan's Ageing, Labour-Starved Construction Industry Pouring Investment into AI, Robots," *The Straits Times*, November 5, 2019, straitstimes.com/business/economy/japans-ageing-labour-starved-construction-industry-pouring-investment-into-ai.

28   "Digitalisation and Energy—Technology Report," IEA, 2017, iea.org/reports/digitalisation-and-energy.

4

PEOPLE-DRIVEN CHANGE

"Change management is at the heart of any successful digital transformation."

DR. JOHN PILLAY, SVP Transformation, Worley

I T IS convenient for oil and gas leaders to treat digitally driven change as principally a technological problem. This makes sense. The industry is asset-intense with relatively few people; the assets drive financing and capital; and, historically at least, information technologies for the industry (such as SCADA systems) were essentially extensions to the assets.

Digital technologies tend to reinforce this perception. AI is relatively new and still developing, blockchain is still nascent in adoption, and automation in transportation is still in testing. At the time of writing, even cloud computing has just 10 percent market share. Industrial automation and digital automation remain separated.

But the deciding element about whether a company has moved beyond *doing* digital to *being* digital is its attention to the people side of digital. Digital technologies routinely interface with humans at some point, in design, deployment, support, and frequently in operations.

Deploying digital solutions starts with understanding the culture of the business. Oil and gas is a very conscientious industry, focused on safety, results, reliability, and cost. "Risk-taking" and "cutting edge" are not phrases that describe oil and gas. Yet these are the words that attract young talent and encourage transformation. How can a culture that encourages some level of innovation and risk-taking be nurtured in an industry that structurally resists change?

Overcoming **change resistance** is a second key signpost of success. Change agents need to confront negative narratives about digital—that it is overhyped nonsense, a temporary fad that can't possibly make good on the promised returns, or simply won't work in a legacy setting. What does change resistance look like, and how can change agents overcome the natural inertia in the industry?

I devoted the better part of thirty years to helping organizations move beyond their status quo, and I have continually encountered change resistance of various kinds and at various levels. I learned to devote much more time and attention to understanding and addressing the emotional side of change in an industry that is more attuned to cool industrial logic.

How humans feel about digital matters greatly. Are they perpetually under the **Sword of Damocles**, to be let go in the face of industry headwinds, or, by becoming educated in digital technologies, to find themselves redundant regardless?

Many in the industry may be contemplating reshaping their careers for a more digital future but are uncertain where to shift. Others may wish to transition to new energy solutions. Finally, some will simply exit the industry.

But for those of you determined to stick with it, perhaps you have identified possible technology changes, framed an exciting new business model, and surfaced the changes you need to make. How do you get your team aligned and ready for a change not only in their business model but also in the tech they use and the culture of your company?

This chapter addresses in detail the people side of digital, beginning with the culture of oil and gas. It is slow-moving, careful, and conscientious, based on the danger of oil and gas. This makes it at its baseline more change-averse than other industries.

The Power of Culture

To succeed on a change initiative, you need to consider the culture of your company. According to Jay Billesberger, "we really focus on the culture, less on the digital side, to get ready for digital that is to come."

Company culture is complex to articulate. It can refer to the interactions between employees in the break room. It can refer to the ways your company handles videoconferencing calls. It can refer to how you view business partners, contractors, and suppliers. It can refer to your values and norms. Understanding the culture is critical to ensuring that change works, because culture is often either a barrier to or enabler of change.

There are multiple ways that company culture can manifest, and each of them will impact change in different ways.

IN APRIL 2015, Shell acquired BG Group, and I had been working separately and in parallel for both organizations prior to the merger. There could not be a greater study in contrasting cultures. Shell's managers expected their consultants to move fluidly through the organization, meeting with other managers and assisting with alignment efforts. BG Group was strictly hierarchical; project messaging was tightly controlled; and even discussing your project work outside your immediate team was grounds for dismissal.

Not appreciating this distinction gave rise to one of my worst dressing downs ever, from a BG executive when I mistakenly briefed his peer about my work—as was totally expected at Shell and expressly forbidden at BG.

The culture of your employees and co-workers can manifest at a micro level. How do people talk to one another? How is news communicated? Do people blindly follow orders, or is there a continuous dialogue of dissent? Does an employee get to set their priorities for the day or are those set by managers? Understanding how the company functions socially goes a long way to gauging how people will react to change.

The Culture of the Industry

Oil and gas is a very conscientious industry. It generally holds itself to the highest in safety standards. Its employees endeavor to meet all of their commitments and are responsible to their shareholders. In many nations, energy is viewed as a social good; the industry may be subsidized by governments to achieve affordability, and, in return, oil and gas plays a social role as a major employer.

The industry makes money only when its assets are active, and it runs 24/7. Any **shut-downs** or turnarounds need to be as quick as possible to maintain cash flow. Assets, being long-life, heavy, and dangerous, need to be kept in working order at all times. Oil and gas rewards reliability.

This conscientiousness also makes the industry risk-averse. For this, we should be grateful. The industry's products are dangerous. Crude oil is toxic, caustic, and explosive. Accidents and leaks can cause widespread ecological and social harm. Energy shortages can result in social unrest and economic damage. When accidents occur, regulators and governments impose stiff new rules. The industry de-risks operations as much as possible to avoid the many negative consequences.

However, this risk-averse culture can result in an aversion to change as well. If the operations are running smoothly, safely, and reliably, then it becomes hard to make a case for change. The industry rightfully has zero tolerance for mishaps. Poorly considered changes can lower reliability, raise costs, and contribute to accidents.

As a result, oil and gas is much more likely to be a fast follower of digital innovations. Smaller companies, which lack resources to work through change, prefer to follow industry leaders such as BP and Shell.

Worries about cyber activity also weigh on the readiness of the industry to address digital change. Facebook's multiple data breaches over the past few years have eroded its brand.[1] If they, a digital giant, can be a victim of data breaches, why not energy companies?

Increased digitalization, even at a smaller scale, can create vulnerabilities that need further development and change to solve. In 2012, Saudi Aramco made the news when one of their facilities was the target of a cyberattack,[2] a situation similar to Colonial Pipeline in 2021. In both instances, the assets had some digital smarts, but the criminals had found ways to exploit the systems in place. The systems were not robust enough to handle cyber, and they lacked the change-flexibility to make improvements fast enough.

To make the case for change in oil and gas, change agents must confront this risk-averse, conscientious culture. Manager of Upstream Digital Heather Wilcott, from Imperial Oil, observes of her company, "We built a culture of processes for safety reasons. We rewarded people for following the processes and adhering to them. And now we want them to try new things."

Company Culture

The culture within an individual company determines its ability to change. Their relative readiness, their talent mix, their organization structure, and their rewards systems all contribute to the capacity to tolerate and manage change.

Scale creates challenges. Different departments may have slightly different metrics, and as a company grows and expands throughout the value chain, the metrics can vary considerably. If there are conflicts between goals and metrics, friction can result. The friction can be immaterial if the company is achieving its own goals, but it can quickly become a problem in times of organizational stress.

Once friction exists between parts of the company, it can become an inhibitor to change initiatives. Departments like things the way they are, and they don't react well when somebody from a different part of the business tries to tell them how to do their job better.

From a governance perspective, setting clear expectations and promoting symmetry across goals and performance metrics can dramatically improve the prospects of introducing change. This is easier in smaller, more focused companies. A small upstream company can better align all of their departments with the goal of delivering two complex horizontal wells; a company that operates a variety of assets in many different countries is going to have a harder time aligning people with a common vision.

I LIKE to distinguish between people who work in the business (the workers) and people who work on the business (the changers). The workers focus on daily operations and execute the necessary routine work that keeps business going. They value stability, reliability, and predictability. The changers focus on inefficiencies and how to correct them, opportunities for growth, and innovation. They value constantly improving results, competitive advantage, and innovation.

Much change resistance occurs when some well-intentioned change to the business meets the beautifully running status quo tended with care by the workers.

Even at the personal level, digital efforts can be thwarted. What your organization tolerates in the form of resistance to change sets the pace for success. As one executive put it: "20 percent are enthusiasts. 60 percent are compliers. Ten percent dig in and resist, and 10 percent actively obstruct change."

Taking on the Culture

As we've established, business culture can create barriers to change. These come from the overall culture of oil and gas, the culture of individual companies, the culture of particular departments and teams, and the cultural relationships created between individuals. Taking these on is important to getting change right. You need to accommodate culture when it is required or change it when necessary.

How do you know if your organization is ready to embark on the changes enabled by digital innovation? Why not collect some data? There's little point in developing an elegant, sophisticated, and disruptive game plan only to discover that the organization doesn't think management is serious, the team lacks the capability to execute, or the changes so threaten the workers that they slow-walk it forward or refuse to go along. As the late Peter Drucker reminded us, culture eats strategy for breakfast. As you work on your digital program, you should develop a sense of how ready your organization is to cope with the outcomes of your changes, and for this you need data on attitudes throughout the organization.

1 **Is there a difference between how managers think about change versus how the troops do?** Sometimes managers, who are measured on high reliability and minimal upsets, can feel pretty ambivalent about introducing digital innovation. It just feels like more work. Sometimes the troops can see change as simply "more work, same pay," or "big brother is watching me" or "automation is after my job." They, too, can be pretty cynical about any kind of change.

2 **What are the views of the corporate office, whose people are possibly more exposed to conversations about digital innovation, versus the field operations?** Sure, there are probably lots of opportunities to apply digital solutions to corporate work, but most (if not all) of the really useful digital advances in oil and gas will be found in the field.

3 **How do different employee groups, such as technical, professional, and management, feel about change?** You might suspect that the technical and professional cadre are the most open and best aware of how digital changes are impacting the industry, but is that true? Frequently their jobs are tightly coupled to specific technologies that they may be loath to replace. What about procurement, who insist that any digital innovation implemented should be in the market already (if it's already in the market, perhaps it's not innovation)?

If there are significant differences in attitudes towards digital across your company, you need to craft some actions to help improve the adoptability of the strategy. But collecting data about attitudes involves reaching people who are widely distributed across big geographies and lots of field offices. The assets run round the clock, which requires multiple shifts of workers to keep the machines running, making it hard to reach everyone via interviews or focus groups. Just getting to the field can be a multi-day journey.

The differences between the city-based back office and remote field office can lead to a lack of unified planning or messages. As Heather Wilcott put it, "I still continually run into worker-level people who have never heard of the stuff that we're doing or their roadmap for their business unit."

A useful technique is to conduct a digital readiness survey of the organization facing change. Survey tools are ubiquitous, readily accessible, inexpensive, and robust. The simian twins, Mailchimp and SurveyMonkey, have cornered the market on engaging with lots of people, so start there.

Your first challenge is the overall design of your survey—how long, how detailed, how much time to complete. Shorter surveys can yield higher response rates, but you collect less data. Set up your survey so that it is "responsive," meaning it works reliably on any phone, tablet, or browser, and can be completed anywhere, anytime.

Here are a few sample questions to consider:

- How familiar are you with the following digital technologies? Phone apps? Cloud? IoT?

- Compared to our competition, how are we doing in exploiting digital advancements?

- What is the potential for digital advancements to improve our business?

- Where do we need to change in order to benefit from digital advancements?

- What priority should be placed on exploring and implementing digital advancements to benefit the business?

These questions will frame how the company views digital and what the overall cultural attitudes towards change are. The more data you collect, the more you can accurately discern the way that digital change will be seen by the company and how you should build your strategy, message, and approach.

Reorganize for Success

When you bring forth digital changes, your people will be wondering about your ultimate intent. This can be positive for people, as their talents and skills are being revitalized by changing their duties, but it can also be disconcerting. New technology often leads to job displacement in addition to job enrichment. For many people, digital innovation is just code for "we're using robots to do your job."

But the fact is this: an innovation that saves 10 percent of ten people's time in a day frees up one person from doing that work, provided you reorganize the work to capture that effort. Plan for the reorganization as part of the adoption of the technology, make plain your intent, and commit to supporting those who are displaced (with a new job elsewhere in the company or with a solid outplacement package).

Digital changes can help you transform roles and duties to make better use of your people's talents and time. Being a better leader and leaning into change will show you where there are gaps and improvements to be made in terms of organization. You also need to be ready to shift gears with agility, as it may not be immediately apparent which roles would fit which people best. As one of my case interviewees put it, managers want employees out fixing wells, not spending hours working on a spreadsheet trying to decide which well is the priority.

In oil and gas, digital is not about letting people go or laying off talent. It is about making the most of all of the people on your team. Above all, you need to stand by the people in your company. You must be behind your team and not simply behind change.

Support means not dressing down a team because a digital trial failed, or a month-end deadline was missed, or a payment went astray, or promised results didn't materialize. Digital leadership means helping people where they need it, taking their criticism and suggestions to account, clearing roadblocks, and talking up results.

This means focusing not just on the technology, nor on the overall benefits to the company. You need to empower people to use the technology.

"Companies have a kind of permafrost in the middle. The boards and executive get the need for change, the frontline workers want to see improvements, but the KPIs, procurement rules, and culture create this frozen middle layer."

AZAD HESSAMODINI, President, Strategy & Development, Wood

Get the Message Right

Digital is often seen as a people replacer. Your culture will be wary of changes that threaten jobs. Many innovations, such as autonomous vehicles, are already being presented as ways to replace people with automation, not improve work or make people better at their jobs. If oil and gas is already change-averse culturally, then anything that appears threatening will be rejected.

The message needs to be right for digital technologies to hold. If you want the culture to accept the introduction of these novel technologies, then it has to be framed in a positive light. Here are some ways to frame the message:

Growth and competitiveness: Emphasize the growth opportunity in digitalization. Frame discussions not just in terms of cost reduction or efficiency gains, but also around how the company can grow through digital innovation.

Job creation and enrichment: Emphasize the creation of new jobs and the improvement of existing ones. Identify specific jobs or roles that can be created with digital.

Safety and security: Emphasize how secure and reliable modern digital technologies have become. Make it clear that the company and people's careers are not under threat. Describe how digital innovations do not imperil the safety culture but serve to enhance it.

Create the Conditions

The next step is to create the conditions to enable the change to be successful.

In general, change is acceptable in oil and gas when it's handled slowly, methodically, and under stringent MOC processes. This approach helps assure that mechanical changes will not cause some catastrophe or create unsafe conditions given the dangers and risks involved in handling hydrocarbons.

Compare the hard technology world to the back office, with its paper documents, Excel spreadsheets, and engineering software. Creating an agile, change-tolerant back office is one area where you can get digital done quickly, while the field takes the time it needs to meet its safety and compliance requirements. Many back-office settings offer ample opportunity for fast, successful change. As Jay Billesberger put it, "You can't fall off the floor."

Once you understand the way people think in your company and how they will potentially respond to change, once you consider their interests and frame the narrative to align with their mindset, you are creating the conditions for digital acceptance. In time, you may even experience a broader cultural shift, assuming you have support from leadership, towards a more open, embracing, and agile business culture.

However, one major hurdle to overcome will be change resistance and its standard bearer: the hedgehog. It will arise in all conditions, no matter how much work you do to align the culture.

Countering Change Resistance

Change resistance is when companies, departments, teams, or individuals refuse to, or are unable to, embrace change initiatives. Sometimes employees don't understand the change being requested. Sometimes they fear the consequences of poorer performance as they adapt. Sometimes they even disagree with it.

Hedgehogs, the Change-Resistance Flag Bearers

One kind of change resister reminds me of the hedgehog, the cute garden critter common to the English countryside. I call change resistant individuals hedgehogs because at the first sign of change, they curl in balls with their objections pointing out, making it hard to engage. A common beach defense to block tanks during war is also called a hedgehog. It's made of steel, designed to be immovable, features lots of sharp pointy edges, and does a lot of damage if you confront it head on.

THE CFO of an oil company was encountering puzzling objections from his managers in the face of digital opportunity. His firm had acquired a large oil asset and the seller had shipped over boxes and boxes of paper files associated with the asset (which was not surprising, as the oil asset had been in business for decades). The managers dismissed his suggestions that the paper be converted to digital to make the files more searchable. "If we want anything from those files, we'll just go looking for it," they claimed.

However, none of the engineers really wanted to devote their time to pawing through dusty papers. It's work better suited for an archivist than an engineer, after all. The CFO suspected that he would eventually be paying for the data and analysis a second time, the old boxes unopened.

In management there are a lot of hedgehogs. Hedgehogs say things like, "This won't work in oil and gas," or "We can do this cheaper manually," or "We are too regulated, too fragmented, too business-to-business," or "Our operations are 24/7 and can't be taken offline except in turnarounds or emergencies." Heather Wilcott describes it as "The Yeah But syndrome. 'Yeah, I could do what you are asking, but what about . . .?'"

For example, brownfield plants and equipment put into service many years ago were never designed for a digital world, do not easily support wireless networks, and are manual, paper-based, mechanical, and human-centered. They are hard to change, and it's not uncommon to find human hedgehogs running this kind of plant. Even after the pandemic effect, and in the face of oil market upheaval, hedgehogs still question whether digital solutions offer any value.

Hedgehogs in Charge
Here's what happens when the hedgehogs are in charge.

Slow pace of change: The pace of change with a hedgehog at the helm will be necessarily slow. All possible contingencies and remote **black swan event**s are to be studied and deeply understood. Hedgehogs surface a continuous stream of new risks, and all are treated with the same level of care regardless of impact.

Poor talent outlook: Hedgehogs that block digital change from their companies are inadvertently laying the groundwork for talent erosion or, worse, zero talent attraction. Employees in oil and gas have been schooled in the discipline of the market through the downturn. They know they are dispensable, and they will be looking to future-proof their own careers. This isn't going to happen if a hedgehog blocks change.

Timid digital strategies: Hedgehogs promote small evolutionary steps, and not big changes. They hold out for slow multi-year roll-outs. With digital innovations impacting all parts of the economy at the same time, and accelerating in the pandemic, boards are instead asking management for a complete digital strategy refresh.

An ecosystem of hedgehogs: Hedgehogs double down on their existing ecosystems of like-minded incumbent suppliers who, like them, are slow to embrace change. But it's the small startups that have the upper hand in embracing digital. They have fewer constraints, faster decision-making, lower initial approval hurdles, and creative talent.

Hedgehog team-think: Hedgehogs will hire other hedgehogs and create a kind of top-to-bottom thinking model that discourages change. This manifests when frontline workers challenge innovations on the basis of safety concerns, even though safety and innovation should not be an either-or decision. Why is digital innovation positioned as anti-safety? Surely, it's safety *and* innovation? This is hedgehog team-think at its best.

I've spent a lot of time with well-meaning hedgehogs in oil and gas who are finely practiced in diplomatic skills. Here are three strategies that they employ to keep digital programs from making any real progress.

THEY'LL TRY TO MANAGE THE CONVERSATION

People who don't like change may initially attempt to control the conversation because talk is cheap.

Any presentations or training about digitalization have to take place on-site, so that the change agent will have to work harder to influence the narrative, promote fresh lines of thinking, and counter-act subtle but dismissive language about digital.

The problem is that many key oil and gas facilities are well away from civilization. Some sites insist that vendors provide briefings for free, at awkward or inconvenient times, which will minimize the number and duration of such sessions, as well as the number and kinds of experts that attend. Lots of questions will go unanswered without the right expertise available.

There is real value in visiting industry leaders to gauge the opportunities presented by digital. However, many oil and gas sites block requests for travel to conferences or to courses on the basis of safety, lost hours, or cost.

At some point you may be given funding to send people to training or conferences or on benchmarking tours. Sites pick the attendees, frequently not the most influential team members. There's always something more important that they can be working on. Instead, employees of little consequence or low influence get the nod. Need-less to say, someone not well regarded by their peers will be unable to convince the team that digital is something to embrace.

THEY'LL SLOW THE PACE

If you've had some success in laying the groundwork, you might dis-cover the pace of change seems oddly slow compared to the pace of change in the digital industry. It turns out that enthusiasm for change is susceptible to delay.

Maintaining pace is hard. Any manager can come up with myriad reasons why timing is bad for your digital efforts. Perhaps some new equipment is being delivered, or some other corporate initiative that must take precedence is occupying people.

A common tactic used to slow down a digital effort is to require compelling evidence that the proposed digital solution is completely risk-free. That means it has five or more customers, all of whom are in the same geography and in precisely the same line of business, and willing to speak openly about their experiences. Good luck.

It's really tough to overcome mythical boogey men whose risks are painted as high-impact and high-probability (cyber threats are the current favorite). Putting all digital changes, regardless of what they are, through the full MOC process will tie them down for months.

The final and frequent digital killer in oil and gas is the lack of network connectivity in the field. Sites are quick to point out how satellite uplinks are too expensive, unreliable, or capacity constrained.

THEY'LL IMPEDE PROGRESS

The third and final way your digital initiative will struggle is during execution. You find that you need access to certain key resources, such as a highly experienced but supremely busy engineer. Negotiating their involvement is challenging because their performance metrics for the year are locked down and roll up to their boss. Freeing them up means some other commitment will not be met. You need a get out of jail free card for them, their boss, and others dependent on that commitment, or they won't be available for the digital project.

Digital is about trying things to see what works and what doesn't, but oil and gas is only about doing what works. Only those in R&D set out to do something that might not work. Workers in oil and gas generally have clear objectives to meet, and they know how to meet those objectives with the tools they have at hand. Getting them to try something different that might not work is not going to be greeted with enthusiasm.

STOPPING DIGITAL projects too quickly is another key action that can plague progress. An oil company started a trial to adopt robotic process automation tools to carry out battery balancing. The bot failed dozens of times before its developers figured out all the nuances involved in this complex but routine task. Having critical help on hand at month end to bring bots back to life after failures was hugely disruptive to the business. Many oil companies simply halt development after just a handful of failed attempts.

Clearing Out the Hedgehogs

Managers know exactly who their hedgehogs are. Hedgehogs dislike all change, not just digital change. Many are successful in their roles because they have rare skills. They are hard to replace. But with the combined pressures of the environment and the capital markets, holding on to them is becoming organizationally costly. Digital technologies are actually helping make them dispensable, and along with the upside offered by digital enablement, now is the time to replace them.

Private equity gets this idea. In 2006, I helped Fortress Investment Group, a private equity firm, absorb Intrawest, a Canadian leisure industry player (golf courses, ski hills). After spending time in Vancouver with Intrawest, I met with Fortress managers in New York to discuss tactics and, in particular, my observations about manager resistance to the changes Fortress wanted. Fortress's philosophy with all of their deals was to try to keep the existing management in place, but after a few months of patient encouragement, if the managers were not yet on board, then they were packaged out.

Learning to Trust Digital

Part of managing change resistance is understanding the question of trust. People have trouble trusting novel technology. This comes from their ignorance over how the technology works, fear for their jobs and their value as an employee, and their current trust in existing technology. For example, I remember watching my dad at work when I was growing up. He was a math whiz and correctly tallied up long columns of numbers in his head. One day the boss issued these new, battery-powered calculators to the troops. It was the coolest technology, with its space-age red symbol readouts, tiny keys, and form factor. The funny thing was, Dad added up the numbers using the calculator, as the boss wanted, but then he redid all the math by hand because he didn't quite trust the calculator, the algorithm of his day, to get it right.

Digital, however, is different. The changes that digital solutions can unlock are more sweeping, more cultural, and more job-changing than almost any technology before it.

The Rise of Machines

Digital changes our relationship with machine tools. Recall my sketching out a framework for thinking about digital innovation:

1   The IoT (sensors) generate torrents of data.

2   Artificial intelligence and its companions, such as analytics and machine learning, ingest, interpret, and analyze that data.

3   Automation (robots) apply that data in the real world to get work done.

4   Cloud computing houses much of the data, intelligence, and robotic controls.

5   Blockchain provides immutable evidence that the sensors are reliable, the AI engines are correct, and the robots haven't gone rogue.

There's a lot of evidence that this framework is solidly grounded in reality:

- Google engineers capture a 40 percent energy saving in their global data centers by letting an AI engine directly manage energy use.[3]

- BP turns over its oil and gas field operations in the US to AI because the AI engines are superior at running oil fields once they learn how to optimize the assets.[4]

- Ambyint captures a sizeable purchase order from Husky to deploy its edge AI devices across an oil field, which will pay back in just two months.[5]

- Suncor and CNRL use robots in their heavy hauling operations.[6]

* Porsche and other automakers put their latest cars on blockchain to create new customer services and lay the foundation for autonomous vehicles.[7]

As it is with digital, none of these technologies is particularly new. The field of artificial intelligence dates back to well before the dawn of computers. The first digital and programmable robot, the Unimate, is from 1954.[8] Online bot technology originated from the online gaming industry a decade ago. Blockchain is based on three technologies (distributed computing, peer-to-peer networking, and encryption), all of which have roots in the earliest computers.

But just reading about them can induce a mild sense of discomfort, that the machines are taking over.

Work Is Evolving

These automation building blocks that are driving changes to work display no prejudice. They're impacting heavy and light industries in equal measure. It's only the pace that differs. Automation is now plainly visible in agriculture, where farmers use drones to supervise planting, irrigation, and pest control. It's in shipping, where unmanned cargo vessels are in field trials. It's in material science, medicine, media, even retail.

The energy sector, in particular, is undergoing a second shift as the world looks to embrace more energy from electrons and less from molecules, particularly in transportation. This shift has huge implications for work. The molecules today that yield energy to us when we transform them (such as combusting gasoline in a car engine) require a healthy level of human supervision to keep the mechanical apparatus that contains those molecules from gumming up, overheating, and wearing out. Electrical gear has many fewer moving parts overall, and a dramatic reduction in the number of parts that need to be cleaned, cooled, and replaced.

The result is we will need far fewer people to maintain this equipment. Fewer people on-site to keep an eye on things, to use tools to measure how equipment is behaving, to lubricate parts, and to detect aberrant sounds or smells. Electrical equipment lends itself to robotics, remote control, and remote monitoring more so than molecule-based machines.

REMOTELY CONTROLLED technology does not always work out in practice. For example, we have all the technology we need for aircraft to fly without onboard pilots: there are now flying drone competitions that demonstrate what can be done. The same is true for submersibles and helicopters. But today, no airline offers flights without a human pilot at the controls at all times, even though it's unnecessary. Pilotless aircraft demonstrate that successful technology must meet the MASA principle—most advanced, but socially acceptable.

Many mission-critical assets, especially in heavy industry, still feature a healthy, but expensive and technically unnecessary, on-site human management team. The social pressure that might block adoption (as with Google Glass) is not present in some obscure plant far from human settlement.

It begs the question why management teams, particularly in the energy sector, facing critical environmental and competitiveness challenges, are resistant to embracing greater levels of automation in anticipation that automation is coming.

The Uncertainties of New Technology

The oil and gas industry has removed much of its uncertainty over the years. Engineers rely on the technology model designed forty years ago, since the development of SCADA systems in the 1980s. Analog sensors are built to known industrial standard, embedded directly in the piece of equipment, and hardwired to a control panel, or to a front-mounted gauge. There are no uncertainties in the industrial standard, nor in the tests of compliance that the devices and gauges must pass. Copper wire is highly robust and the SCADA system has been working reliably for decades now. The data is extracted from the historian directly into an Excel spreadsheet onto a trusted employee's computer for analysis and interpretation. The people involved are known quantities.

The engineers know how all this stuff works. It's been part of the curricula and disciplines for decades. Older engineers pass this knowledge on to new recruits.

New technologies introduce fresh uncertainties into this stable world. Imagine strapping a new modern wireless sensor on a piece of equipment, using an AI engine on the cloud to interpret that data, and making decisions based on the results. Here are eight uncertainties:

1 the sensor itself, its technical features, and its compliance with industrial standards;

2 the sensor mounting and how reliable it is to capture data correctly;

3 the power supply to the sensor and its reliability (AA batteries don't make the grade);

4 the data the sensor generates and the potential for compromise (an embedded sensor cannot be as easily compromised as a strap-on wireless);

5 the integrity of the wireless network that moves the data from the sensor to the cloud analytics engine;

6 the technical competence of the algorithm's author;

7 the integrity of algorithm itself; and

8 the results the algorithm generates.

AI proponents are frustrated with the slow pace of adoption of AI technology in oil and gas, but many offer little by way of helpful response to the uncertainties above, and some social norms compel them to block their own success. For example, they cannot, or will not, clearly explain how their AI algorithm technically works, perhaps out of a concern that their IP will be compromised, which means the engineer has to trust it.

AI in the Field

Let's assume that some brave oil and gas company kicks off an AI initiative, rolls a shiny new algorithm out to the field, and hopes for results. Assume that they resolve the problems of sensor provenance, network reliability, power supply, and connectivity. Remember, the field engineer does not understand how the algorithm works. Soon the algorithm starts generating analysis.

Consider four scenarios:

We're both right: The algorithm correctly interprets the data, and the interpretation matches what the engineer thinks should be the result. The AI machine is improved, but he's irritated that the company spent a lot of money on something that he could already do. The full benefits of a smarter machine are deferred.

You're right, I'm wrong: The algorithm correctly interprets the data, but the interpretation differs from what the engineer thinks should be the result. Now the engineer faces a dilemma: Is this a false positive? Does he take the recommended action for an uncertain outcome, or ignore the recommended action and rely on his intuition or a sidebar manual exercise? Performance metrics and targets compel the engineer to weigh the business and personal consequences to determine the safe path.

What if the machine is correct and proves the engineer has been wrong all this time? Will there be repercussions? Who wants that embarrassment?

Our engineer reverts to previous analysis and ignores the machine. He loses a learning opportunity for both the human and the machine, and he'll spend time and money trying to replicate the algorithm. At least he can claim to have avoided a possible catastrophe, which he would have avoided anyway, and he won't be embarrassed by a machine.

But if he follows the machine's recommendations, he is better off in both the short term (because of a smarter decision and fewer wasted resources) and in the long run (because the machine is made smarter).

You're wrong, I'm right: The algorithm incorrectly interprets the data, and the engineer agrees that the interpretation is incorrect, forcing the engineer to rely on previous know-how and analysis. He's irritated that the company spent a lot of money on something that doesn't work and he's no better off. His choices then inform the algorithm, which gets a bit smarter and might pay off later on, but at the risk of doing himself out of a job.

I can't tell: The algorithm incorrectly interprets the data, but the engineer can't tell that the interpretation is incorrect and has no better analysis to leverage. Again, the engineer is in a dilemma: What if the machine is wrong? If she follows its recommendations and it doesn't work, is she to blame?

She's in a bind, and has no option but to execute, and there's a failure. She'll wear this at the next performance review, but at least she can cast some shade on the algorithm. The algorithm is a little smarter.

The above examples illustrate the problem that much of oil and gas faces as it confronts these new technologies. Why don't we trust machines to perform mission-critical tasks (even for tasks that machines clearly do better)?

There may be several reasons:

- Building trust within an interactive and mutually dependent relationship with other humans is something we are just wired to do. We don't have that same relationship with a machine. Machines lack the human-type needs that drive their decisions, and so we don't feel part of that decision-making process.

- For the most part, we care about being reliable to those that depend upon us. Machines don't care about how they are perceived by humans—they just follow their instructions. The developers of those machines are disincentivized from being transparent about how their complex and proprietary algorithms behave.

- We like that humans are adaptable and can problem-solve in the moment when things go wrong (as with our pilots). Machines are getting much better at this, thanks to fleet learning, but we still aren't confident enough in their abilities.

- We like that humans can exercise judgment, such as when to break a rule or create something novel. Machines are bound to their instructions, and we're uncomfortable with the idea that a machine could break rules.

- We like that humans can be held accountable for their actions and react well to rewards and punishments. We have lengthy (albeit imperfect) experience levying consequences against other humans that stretches back through the whole history of humankind, and it is central to the experience of being human. We are more comfortable having a human "at the switch" (or at least in the loop, keeping watch over the machines and having supervisory control).

- Machines can be programmed, using gamification theory, to respond to rewards and to avoid punishments, but only in a math sense, and not an emotional one. Shun a human for a transgression and their behavior might change; but try shunning your digital home assistant—Alexa will just ignore you back.

The Five Components of Trust

If we are to attain the highest performance out of our relationships with smart machines, we need trust.

To gain trust, especially in an industrial setting, there are some very specific things that need to be overcome:

1 **Lack of performance:** Is the machine even going to work? Will it do what is needed?

2 **Lack of long-term track record:** The machine may have performed acceptably in the demo or pilot test, but how do I know it

will work over the long term, under every conceivable condition (cold and hot weather extremes, loss of power or communication, unexpected operating conditions, emergencies and exceptions)? Can I rely on it when the context completely collapses, and creative problem-solving and adaptability are needed?

3 **Lack of integrity:** Machines lack morals (unless we program those in) so how do I know that the machine will be truthful with me? After all, we now know that existing algorithms (or their designers) aren't always truthful and forthright about data—consider the numerous scandals associated with the social media giants. Or, more concerning in an industrial setting where safety really matters, how will the machine handle conflicting decision-making criteria, such as the famous trolley problem that self-driving cars must address?[9]

4 **Lack of clarity:** Why does the machine make the decisions it makes? Under what conditions will those decisions change? How can I influence those decisions in real time as my conditions or drivers change (for example, oil prices change or someone extends an offer to buy my company)?

5 **Lack of accountability:** If the machine makes a poor decision, how do I hold it accountable? Can I give it consequences that matter? Also, in complex industrial settings, many small decisions by many participants add up to larger outcomes—how do I separate out the contribution of the machine's decisions relative to decisions that others made?

IF MY self-driving car gets a speeding ticket because I stipulated conditions that caused it to make poor decisions, I should pay the ticket. However, if the software was written in a way that allowed it to speed of its own volition, then the software vendor should pay.

If my AI drilling management program achieves better-than-expected financial returns due to new upgrades from the vendor, does our contract discuss how these additional benefits will be shared?

These are all critical questions—how do we answer them? To achieve high performance in our relationships with smart machines, we need to be able to trust them, which begins with the five components of trust:

1 **Competence:** Do I believe that all the participants in the relationship have sufficient capabilities to fulfill their respective responsibilities?

2 **Reliability:** Do I believe that all participants will fulfill those responsibilities reliably, on a consistent basis?

3 **Transparency:** Do I believe that I am sufficiently aware of what the other participants in the relationship are working towards? Would they make decisions that are compatible with the decisions that I will make? Can I influence their decisions? Do they keep me in the loop?

4 **Aligned integrity:** Do I believe that the other participants are always being truthful with me and adhering to our agreed-upon principles or norms of behavior? Are we operating from compatible sets of priorities? Put another way, do we have shared values, morals, and ethics?

5 **Aligned accountability and motives:** Do I believe that the other participants want the same outcomes I do? Is there an alignment of interests among all of us? Are there meaningful motivators (bonuses) or deterrents (consequences) to help maintain and nurture this alignment?

Note that these components involve not only *fact* (such as, is the other party competent?) but also *perception* (as in, do I believe that the other party is competent?). It does no good if there is a mismatch between fact and perception.

If the other party is competent (or reliable, or so on) but I harbor some doubts, then there is no trust. Similarly, if I believe that these attributes (competence, etc.) exist when they actually don't, eventually the truth will come out, and the mismatch will result in broken trust.

Not only must these five components be present, but we also need to make sure that all the participants in the relationship are aware of them and that everyone authentically knows that these are present—easier said than done.

Also note that *all five* must be present at *all times,* or trust is compromised. Many of our interactions take a "grow in trust" approach, where trust is built over time as evidence mounts that these components are in place. When even one of these components goes missing, even temporarily, then things take a step backward and trust needs to be rebuilt.

Trust needs to be both earned *and* maintained. All of us have experienced that this takes work and commitment.

Fortunately, behind every machine is a team of people (designer, programmer, operator, and more) at least some of the time—building and maintaining trust really needs to happen with them. As users of smart machines, we need to make it very clear to the technology designers and vendors that we expect these five components to be addressed in a concrete manner. Call it the social contract for digital innovation.

Maintaining Trust in a Smart Machine

To earn and maintain trust in their smart machines, the designers and vendors of digital technology need to address very explicitly the five components of trust.

1   **Competence:** This is the easiest aspect for vendors to prove—simply demonstrate that the products (smart machines, algorithms) achieve the promised performance under test conditions.

2   **Reliability:** This one is harder to prove, because now performance assurances need to cover off not just "easy" or "normal" operating conditions, but also all the various extremes that might be encountered (the "edge cases"). Also, consistency of reliability can be very situation-specific—my standards for consistency might be very different for a personal navigation and traffic app (80 percent reliability might be good enough) versus the navigation and flight control software on my favorite airline (it better be greater than 99.99 percent!).

Perhaps the solution is to allow humans to intervene in the decision-making process if needed, so that humans, who are better at creative problem-solving, can address the edge cases. However, this doesn't mean that humans need to always occupy every single pilot seat—just look at how your local supermarket handles this at the self-check-out lanes—one human teller oversees six to eight tills, intervening only when an issue comes up. That represents a greater-than-80-percent improvement in efficiency.

3   **Transparency:** As we've previously pointed out, AI developers need to get over themselves and start sharing how their software works, in ways that are understandable to the users.

4   **Aligned integrity:** Settling on aligned principles of behavior can be challenging, especially for some of the edge cases. However, if developers don't significantly open up the conversation to customers on how they approach this, then strong trust will continue to elude these relationships.

5   **Aligned accountability and motives:** Giving a meaningful consequence to a machine is difficult. However, once there is agreement on the outcomes that the smart machine is working towards, plus alignment on decision-making principles and how they will be achieved, then accountability can flow through those agreements to the smart machine's human supervisors.

This approach, which incorporates a human touch to smart machines, provides a symmetry of deterrence and incentives that not only encourages reliability and spurs innovation, but also fits with how humans are wired and motivated. As we deploy more and more smart technologies, self-correcting systems, and autonomous equipment, we need to change our trust equation with technology. Successful change leaders will anticipate that the trust equation is changing because of digital technologies and will incorporate tactics to address trust issues regarding technology.

Digital Leadership

With all the buzz about digital's impact on the oil and gas industry, the demand for digital leadership should be strong. But in what shape is the talent pipeline for leaders in the world of digital oil and gas?

I KNOW personally of at least one chief information officer (CIO) who eventually became the CEO of a large oil company (Tim Hearn, former CEO of Imperial Oil).

To run a big oil company, executives need exposure to the various functions of finding, producing, refining, and selling petroleum. Upwardly mobile executives in oil companies change roles frequently (every three to four years, perhaps) to gain that experience. Oil companies benefit by maintaining a healthy pool of leadership talent from which to draw. It's actually not that uncommon, at least in large oil companies, for the pathway to the corner office to include a tour of duty through the overhead functions, including finance and IT.

Building the Talent Pipeline

My conversations about digital innovation in oil and gas inevitably start with painting the big digital picture of the changes coming to the industry. Many oil and gas professionals still seek a primer on what digital is—definitions, basic concepts, examples, and realized results.

This conversation is very useful, and I don't mind having it at all. Until there's a minimum common understanding of terminology, change drivers, key technologies, and the business case for adoption, the talent question cannot be thoughtfully addressed. Talent questions exist because there's a job to be done, and that job needs to be defined.

But perhaps an hour into it and the conversation inevitably turns to the talent question. It is clear to most executives that digital change is going to be big (it already is); they don't really know what talent they need, they're unsure what talent they have or where that talent is, but they're pretty sure their organizations are short, including within their own IT organizations. As Patrick Elliott notes, "The traditional role of IT is inherently defensive—keep systems running, protect from cyberattacks, maintain uptime. There's not much left over for growth and innovation."

Executives complain that there are not enough data people. Few digital natives occupy leadership roles. Companies have few directors, if any, that have experience with AI, the IoT, robots, and cloud computing. At best, they point to an underwhelming portfolio of small science projects spending minimal capital, but without much oversight and with little apparent impact.

Of course, given enough time, a digital native will eventually rise to a place of leadership in oil and gas, provided they already have the other ingredients that bias promotion in the industry—a background in engineering and a meandering career trajectory that takes them into roles in the various oil and gas domains. But I don't believe the industry has the luxury of time on its side.

The culture of oil and gas poses a critical impairment to developing digital leadership. Really successful rising executives in the industry are disincentivized to invest in new areas like digital innovation. Spending precious career capital on novel, risky, and failure-prone technologies might leave a black mark on a résumé and block a promotion. As an executive it's easy to convince yourself that you're not qualified to work in digital areas and that digital is actually just IT. With all the challenges facing the industry, from intense competition, carbon pressures, and price volatility, there are ample opportunities for career development besides digital technologies.

I can't see a long line of executive leaders standing outside the CEO's door begging for a shot at leading a digital initiative. Better and safer to run that new oil facility or build the next plant.

My conclusion is that there will be a prolonged and systemic shortage of oil and executives with experience in digital innovation and transformation. Over time, this shortage will result in higher demand for the handful of successful executives, keeping the pool of experienced digital leadership small.

And digital is impacting all industries at the same time, leading to the digital leadership vacancy being handed around from company to company.

The challenge is pretty clear—cultural disincentives, a thin pipeline, competition for talent inside and outside the industry, confusion about the need, and a perception of high risk. This isn't the first time that oil and gas has had to confront leadership challenges. In fact, commodity industries go through frequent booms that strain talent pipelines, and the industry has a handful of tactics that it can execute.

Rotate Talent

In March 2019, the big cloud companies (Google, Amazon, Microsoft) descended on CERAWeek to pitch their cloud and AI offerings to the industry.[10] Big digital has discovered the magnificent profitability, cash flow, data intensity, and global scale of big oil. Undoubtedly, they have done the math to conclude that it will take the big balance sheets of oil companies to invest the sums necessary to pay for decarbonization of the energy sector, a sector that they are also targeting through autonomous transportation.

But it's unlikely that the big digital firms have anything close to the expertise needed to understand the business challenges of big oil. Digital sells to many industries, and big oil hasn't been a serious buyer to date.

I suspect that the digital outfits would welcome the opportunity to rotate some of their talent into oil and gas and, in exchange, rotate some oil and gas talent into a digital company. Amazon would benefit from exposure to the challenges of enabling digital smarts in field assets and legacy SCADA systems, and oil companies would gain from learning how Amazon manages to release new software every few seconds.

Tap the Ecosystem

Just about every town and city has now stood up one, or more likely several, incubators, accelerators, co-working spaces, maker studios, rainforests, mash-ups, communities of interest, development labs, field trial zones, and founders' clubs. These innovative work models are full of talent that are aiming to solve the world's problems. They're where you can find the startup community of tech inventors and innovators.

They are also a fertile place to recruit for talent because startups are demanding. The pace is unrelenting, the uncertainty is draining, and the success rate is low. To quote Gimli in *The Lord of the Rings*: "Certainty of death. Small chance of success. What are we waiting for?"[11]

It is a certain that some bright sparks will have joined a startup and become disillusioned with its direction, leadership, unpredictability, or compensation. Some of the leaders will have cashed out and will be looking for new horizons and new opportunities to exploit.

Look Outside

When the price of oil rocketed to north of $100 a barrel, the imperative to grow was overwhelming, but the leadership pipeline was again found to be too thin. Big oil relaxed its normal bias to hire strictly on experience, and it recruited from any industry on the basis of merit. Digital has created these same conditions, and it is now time to look outside the usual boundaries of the industry.

But to which industries? I would take a close look at the mining sector, as some miners are well along in rethinking their use of digital innovation to create the mine of the future. The big digital companies themselves are candidate-recruiting territory for obvious reasons. Clean tech has the dual advantage in that many such firms are already smart on the vagaries of the energy industry and incorporate digital solutions into their business models.

I would not overlook consumer-facing businesses, particularly airlines that have thoroughly de-manned their front offices (the modern traveler has unwittingly become ticket agent, check-in desk, and baggage handler); pizza companies that have robotized their operations; and logistics companies, like FedEx, that are practically computer companies.

Lastly, but importantly, I would take a close look at advanced manufacturing. Oil and gas work, particularly in the upstream, is looking ever more like some kind of manufacturing business and less like a prospector randomly searching for treasure. Pull up Google Earth and zero in on the desert surrounding Midland, Texas, for an eye-popping visual of oil and gas on a manufacturing scale.

Essential Qualities in Change Agents

Your organization will have one or several individuals tasked with helping your company adopt digital innovations. Change agents can have any of a number of business titles, but in the main their job is to create the conditions for a business unit or team to adopt some digital innovation. They rarely have the authority to drive change directly into the business—that job belongs to the business unit leader or functional head whom they are trying to influence.

Change agents deal with business culture challenges. They need to address the concerns of people within the business. They have to tackle the hedgehogs. Good change agents are passionate about improvement and the achievement of a better outcome.

In the main, change agents in oil and gas are

- equipped with an engineering education;

- successful problem solvers from various domains, ranging from field operations, plant management, and supply chain; and

- mature workers, and unlikely to be digital natives.

In other words, they look exactly like me (minus the engineering degree): technocratic, analytic, and experienced.

Change agents quickly grasp why change is so hard in oil and gas. For one, digital innovations from other industries generally don't have to contend with the safety, reliability, and all-weather performance needs of the oil and gas industry. The value of digital innovation is much harder to compute with the level of accuracy that is customary in an engineering context, making funding approvals harder to obtain. Digitally driven change looks very much like it attacks the prevailing business culture, which is much harder to tackle.

The better change agents tend to share the following qualities.

Well-connected, respected, and influential: Making change happen means being able to draw on personal and business relationships, since the change agent will not have the authority to simply command change to happen.

Patient: Changing people's minds takes a lot of time and energy, and digital is changing so quickly that digital adoption is more of a journey than a destination.

Good communication skills: Change agents are constantly communicating their ideas, vision for the organization, impacts of change, and the benefits of change to a wide range of audiences. They need to be authentic, persuasive, and convincing.

Highly empathetic: Good change agents are sensitive to how their message is being received and the impact they have on others. Change can be very threatening to many individuals.

Relentless: Digital is never really "done," in that the technologies and business models are constantly evolving. Change agents are relentless in driving digital change forward. Developing digital is often more trial and error, and good change agents won't quit too quickly in the face of adversity.

Holistic: The better opportunities in digital are cross-cutting and multi-disciplinary, and the new business models that are unlocked tend to favor change agents who are broad-minded.

Business-focused: Technology-led change tends to fail when it is not precisely aimed at a business problem. Starting with the business first allows change agents to target solutions more precisely.

Dissatisfied: Good change agents are not satisfied with the status quo. They tend to be motivated to pursue continuous improvement.

The Change Agent Playbook
The shrewd digital project leader will carefully consider the human change element of the digital project. Rare is the discussion about

what the world looks like on the other side of the fence—that is, to the employees in the business who are expected to embrace digital innovations with enthusiasm.

Lock in sponsorship and support: Successful change agents tie their efforts into the top of their organizations. Digital efforts follow a trial-and-error model, and the high failure rate requires strong executive support. If the troops on the frontline don't think they have management support, they're not going to invest much time in new technology.

Frame the narrative: The best change agents create compelling communications or stories that answer questions from the field:

- What is the vision for the change? What is the compelling picture of the transformed business that lets others embrace the change?

- What's in it for the impacted? What positive benefits are tied to the success of the digital initiatives? How will productivity gains turn into personal growth and not job loss?

- How will the digital change journey be different? How will "fail fast" work in practice, and who will have the employee's back when deadlines are missed in favor of a digital trial?

Boost education: It's hard enough trying to drive change, and harder still when the employee base lags in understanding about the pace and extent of industrial innovation. Good change agents invest in educating the employee base, by whatever means possible, to help minimize resistance to change.

Tune the performance metrics: Change agents find it much easier to drive digital into the business when business leaders' efforts and results are measured. The measures do not have to be complex. Just requesting that managers do one digital thing is sufficient to get their attention.

Run a portfolio: Unlike in the era of widespread ERP adoption, whose impacts tended to overwhelm all other projects, digitalization is a series of smaller changes. Good change agents run a portfolio of trials, proofs of concept, and scale-ups.

Communicate the results: Good change agents share their successes and failures frequently and widely. As one executive told me, "You know your transformation journey is having an impact when people call you up and ask for a presentation based on what they're hearing on the street."

Helping Employees Cope with Change

As much as the energy industry sees itself through a hard engineering lens, it still employs people who need help in dealing with some uncomfortable truths and soft feelings about change.

What Are They Experiencing?

Consider how an employee in a fossil fuel company is likely feeling about the industry.

They probably have shares in companies in the industry, most likely their own, and those shares may well be underwater. Hopefully their pension is more diversified. They may be worried about their retirement security.

They may even have stock options in their company, and unless the options are very aggressively priced, they look like they will never be in the money.

The value of their house, their biggest single asset, is suspect. If they own a dwelling, and they live in the many one-industry oil and gas towns common to Texas, the Dakotas, and Western Canada, there's a good chance that the value of the house is actually tied to the fortunes of the industry.[12] They may be alarmed that the house has seen a decline in value or, worse, might not sell at all.

They all know someone, usually a friend, sibling, or close colleague, who was laid off in the past five years or is facing discharge any day now, with little sign those jobs are coming back. And a few of those retrenchments were due to some digital innovation.

Their kids bring home anti–fossil fuel messages imported into the school system. Their kids might even have participated in a climate strike or walkout. Their kids may not look up to them with the same admiration and respect as before.

Managers have historically said that safety, reliability, cost reduction, and operational excellence are the most important things about the industry, and there's little money or time for modernization. That doesn't quite square with making change happen. Cory Bergh put it this way: "People are hired to do a job. Most do not have a social contract to reinvent their jobs too."

IMAGINE THE inner dialogue taking place in the heads of the workforce:

- I've been doing basically the same job for two decades, and there's been little real change. The documents I use to do my job are still in paper format.

- I never learned to program and don't understand this digital stuff. Sure, I own a smartphone, but that's as "digital" as I get.

- My company is "doing stuff" with digital change, but I'm on the sidelines. Until digital innovations start to scale up, I'm probably not even going to know much about them.

- Am I too old to learn this stuff?

- I'm so close to retirement, why should I take the risk?

- Why should I care if young people don't want to work here?

- I know exactly how to execute my work today and hit all my targets. Why would I put that in jeopardy?

- Where will I get the time to learn the new stuff? With all the departures, I'm carrying more than a full load.

- What if I make a mistake? What bad things will happen to me? Who has my back?

- Where's the prize here?

- I'm getting left behind in a company that's getting left behind in industry that's getting left behind.

- I'm feeling abandoned, worried for my future, lost with the change, concerned about my retirement, paralyzed by the threat of losing my job, and ingratitude from society, which benefits from my work but doesn't have my back.

- My company isn't prepared to even discuss this because, after all, oil and gas doesn't have feelings, just math.

Leveraging Human Emotional Drivers

The very best change agents look at the world through the lens of the employees in their organization and ask if their approach to change is sufficiently sensitive to these people issues. Let's peel this apart to expose the three very human underlying drivers of greed, fear, and pride that make this such a potent framing.

Greed: People figure out how to make an existing system work to their advantage and maximize their benefit. They know the levers to pull, how they interrelate, where to hide their mistakes, how to pump up the results. Sandbagging is an artform everywhere. And when things go sideways, they know exactly who to blame. If they are successful, they have no incentive to change.

Fear: Many people are too close to their next promotion or, more typically in oil and gas, retirement, and they're not going to lose their bonus over something they don't understand. If something goes wrong, they might not be able to fix it, or they might not have time to fix it, and they're not sure who to blame. And if it does work, it exposes the "well-run" machine that they've operated for years as less than perfect. It might even do them out of a job.

Pride: People are justly proud of their well-deserved reputations as top operators. But if they're honest, many people don't even know how to use their own smartphones. There is little chance they're going to put that digital ignorance on display at work, a surefire way to get pushed out the door.

The challenge in the industry today is that the outside uncontrollable forces opposed to the industry (from capital markets and ESG pressures) are creating an overwhelming sense of fear. Uncertainty about how digital will impact jobs is also making people fearful for their very livelihoods.

To counteract these negative forces (fear), good digital programs reframe the narrative to highlight the positives (greed):

- better jobs with less drudgery;
- better career-life balance enabled by regularly working from home;
- career protection from skills obsolescence;
- improved competitiveness, which translates to growth, promotions, and better pay.

One of the case companies even tied the results from their digital efforts into the short-term bonus program, which meant that employees could see just how much impact digital was having on their direct compensation (greed). Needless to say, the interest in digital began to ramp up.

Change agents also know that those less keen on change will present good reasons to maintain the status quo, and those reasons are mostly based on fear—fear that safety will be imperiled, operational reliability will suffer, deadlines will be missed, and, once automated, no one will understand the work anymore. Appreciate what you're up against and be ready for the discussions to come.

Leveraging pride is more of a challenge for a proud industry. The sector has a long track record in running reliably and safely, albeit punctuated with occasional mishap. It takes years of training and experience to become proficient in almost any aspect of the industry. The industry can be immensely profitable.

At the company level, pride translates to brand. Many energy companies now strive for carbon neutrality, and one of the case companies is positioning itself as the "most digital" of its peers. Clever change agents will increasingly craft their communications along the lines of creating industry leadership in digital.

The Stages of Digital Adoption

The job of the change agent is to help as many employees as possible as quickly as possible through the various stages of digital adoption, as illustrated.

Moving beyond any given level is not guaranteed; each individual advances at their own pace, and their progress is influenced both positively and negatively by outside forces as well as internal influencers.

The stages can be defined as follows.

- **Uninformed:** The employee is unfamiliar with the company's stated digital goals and agenda. They have not been exposed to any meaningful communications about the directions the company is taking, the rationale, the impacts, or the expectations.

- **Doubter:** Virtually all employees are doubters first because the status quo works, the company has been reinforcing compliance with the status quo, and new digital technologies are unproven in context.

- **Explorer:** Many junior employees and some experienced staff become curious about the new innovations and seek out new information. They attend lunch and learns, webinars, and presentations. They may still be doubters.

- **Experimenter:** The experimenter is a leader or manager who is willing to put together a trial, or a proof of concept, and encourage their team of the explorers and doubters to participate in an experiment. The experimenter is not yet committed but is willing to learn in context.

DIGITAL ADOPTION STAGES

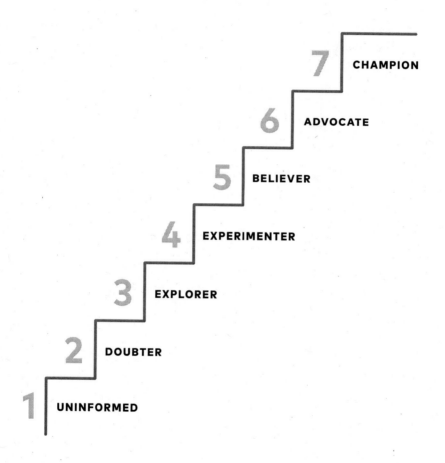

7 CHAMPION

6 ADVOCATE

5 BELIEVER

4 EXPERIMENTER

3 EXPLORER

2 DOUBTER

1 UNINFORMED

- **Believer:** The believer has participated in a successful experiment and now accepts the potential for digital innovations to achieve their projected benefits. The believer willingly engages with new experiments and trials and leads adoption. Successful change agents are attempting to build legions of believers who will drive changes forward.

- **Advocate:** The advocate is a believer who also encourages others (peers) to move through the stages of adoption. The advocate helps accelerate adoption.

- **Champion:** The champion assumes the leadership role for the digital adoption program within a business unit or team.

The case studies in Chapter 5 set out some of the tactics that the change champions have found to be successful.

Helping Employees with a Career Transition

Digital changes drive job changes. Many job categories in oil and gas are vulnerable to disruption by digital technologies. Some jobs will simply disappear. Some jobs will change dramatically. New jobs will be created particularly in support of digital tools. We can anticipate that many oil and gas professionals will be impacted.

DIGITAL SOLUTIONS will trigger painful adjustments in oil and gas among the privileged professional jobs:

- Equipment operation such as trucking, cranes, and construction becomes robotic and autonomous.[13]

- Data interpretation in areas like seismic, reservoir, and production is enabled by AI.[14]

- Facilities management activities like site visits, inspections, or routine maintenance leverage visual sensors to eliminate driving around.[15]

- Back-office tasks such as production accounting and finance switch to AI and blockchain.[16]

- Product trading internationally, including purchase, sale, charters, shipping, and receiving, moves onto blockchain.

Anyone in oil and gas whose job still looks largely like it did in 2015 may be in for a rude awakening.

Companies have a moral obligation to help their employees adapt to digital changes. This can be done by supporting and training employees and making use of technologies that enhance human capacity rather than replace the workers.

There are many ways of making people better and multiple approaches to making digital a solution rather than a problem for oil and gas workers.

Blue Ocean Strategy

If you haven't read it, pick up *Blue Ocean Strategy* by W. Chan Kim and Renée Mauborgne of **INSEAD**.[17] The basic premise of the book is that businesses, like sailboats, should seek to avoid crowded and competitive oceans, red with metaphorical blood in the water from cutthroat pricing and razor thin margins. Instead, businesses should sail for blue oceans, where there are few competitors and space for expansions is plentiful. It's a good way to think about repositioning your career as this wave of digitalization in the industry unfolds. Sail your career to the digital blue ocean.

Where is the digital oil and gas ocean blue?

Data specialization: The amount of data thrown off by sensors is so voluminous that it overwhelms the tools (Excel) that oil and gas has used to manage historically. Only the new and modern tools like AI and machine learning have the capability to ingest, fix, process, and interpret all the data. Data science skills will become part of many different jobs.

Agile methods and process change: The rethinking of work and workflows requires individuals who understand the tools and methods of the digital industry. There will always be demands for some coding skills because the legacy technologies in the industry last for generations. However, there will certainly be growth for those with proficiency in agile methods, UX, and design thinking to take on the task of work reinvention, enabled by technology options such as low- and no-code tools.

Technology support: As tightly regulated operations, oil and gas businesses must invest in the maintenance of their adopted technologies. Consequently, there will be ongoing demand for a range of technology support and know-how around the new digital tools. Cloud computing, augmented reality, blockchain, the IIoT, cyber, and process automation tools are already driving demands for support skills. These technologies are all growing much faster than the oil and gas industry and are in high demand in many industries (agriculture, mining, utilities, logistics, manufacturing) at the same time.

These three areas suggest a growing scarcity of a broad range of skills, leading to high-paying jobs for those with the skills. By marrying deep know-how and skill in oil and gas with those scarce digital skills, an industry professional creates a unique and very rare combination. The education systems are not (yet) turning out professionals in the petroleum sector with digital skills like agile work practices, coding, and data science. It takes time to find the teachers, revamp the curriculum, and obtain the accreditation. But schools are already working on this problem to deliver these new skill sets to the market.

Junior petroleum engineers coming out of school today are not only digital natives (having come of age surrounded by digital technologies), but also able to code and can apply that digital experience to oil and gas.

Be the Career You Want

So how do you go about transitioning your career so you're better positioned for a digital future?

The first step is to pick a blue ocean. It may not matter which digital field becomes your focus area. Just pick one, as they're all in demand and growing at exceptionally strong rates. Get involved in anything related to clean tech, digital, or sustainability. That's where the demand is. For example, one of the projected hot areas for 2022 is data science. Become a local digital oil and gas authority in your field:

- Take a course or two. The university and college systems are all experimenting with new offerings. Or be digital from the start and go the online route in your studies.

- Download freeware versions of digital technology and apply it to your domain area. Nothing speaks louder to an employer when you can say "I've actually coded a smart contract on blockchain."

- Revamp your online presence and résumé to emphasize that you're not just an oil and gas professional, but a *digital* oil and gas professional.

- Write an article or two that showcases how you would bring digital ideas into your oil and gas domain. Publishing your views helps improve your credibility, and having a point of view gives you something interesting to discuss.

- Visit the digital startups in oil and gas. Most oil and gas towns have a thriving sector in digital innovation, and they're always looking for fresh talent to fuel their growth. Perhaps your next job is with a technology company rather than an oil producer.

- Volunteer at the incubators and accelerators in your town. Accidental collisions are one of Silicon Valley's secrets of success.

When All Else Fails

As will happen to a great many professionals in the industry, you may lose your oil and gas job. What should you do now?

Remind yourself that job cuts in oil and gas are usually a response to supply and demand misalignment in the industry. It's not about you, as the first chapter in this book sets out. Capital is scarce, demand is uncertain, and supply is strong.

There are no guarantees, but history has shown that cutbacks in capital spending in oil and gas eventually choke off the supply of fresh oil and gas. Once this happens, the demand for oil and gas exceeds supply and the prices come back up. Companies then restart their capital spending, and the cycle starts anew.

Accordingly, many choose to wait it out because jobs in the industry pay well. When the cycle picks up, your working experience with an oil company or one of the big suppliers is valuable. It signals a level of experience with large and complex company operations, and probably exposure to some of the most sophisticated systems anywhere. The industry likes to buy experience.

But what else can you do?

Look for growth: Regardless of the commodity cycle, some basins around the planet will continue to grow. Basins where the product has been pre-sold or the infrastructure is a large fixed cost are often attractive. Gas supply for LNG is an example, as the gas has been fully contracted. Existing large basins with lots of installed capital, like Canada's oil sands, the Middle East, and the offshore industry, will continue to spend.

Become an entrepreneur: It's not for everyone, but consider an entrepreneurial direction. Oil companies can be accidental incubators for innovation and experimentation, but they are generally not good at commercialization. Perhaps there's some technology or solution that has become stranded in oil company hands, owing to capital constraints or internal red tape, that could become a platform for a commercial advancement. Look for something edgy with external appeal (that is, not specific to a single company) that will be in demand when prices come back. Clean technology and creative digital solutions are candidate areas.

Rebadge as a consultant or contractor: Contracting your time and effort back to the former employer sometimes works. Jobs may go, but some of the work will not, and operators will need to contract it out to get it done. The big consulting houses generally operate under the same logic—that the work is still there—in such areas as analytics, logistics, asset management, and supply chain.

Get into operations or production: Pay a call on operations. Project work in oil and gas may be more exciting, with substantial budgets and tight timelines. But when there are no projects, operations is cash flow, and in tough times, cash is king.

Try the technology sector: Have a look at the tech sector, particularly the very large companies that sell cost or productivity solutions to oil and gas. These outfits generally have solid technical chops but lack oil and gas domain expertise. The same applies to startups focused on oil and gas. They may be great at AI, but they likely know little about petroleum.

Retrain for the future: You might consider investing in training in some of the hot areas, like data science, machine learning, cloud computing, and blockchain. These fields are growing exponentially, whereas the expertise to work with them is not. Plus, the combination of your oil and gas know-how and a new digital tool will be unique in the marketplace.

Focus on cost reduction: The only sustainable advantage in commodity industries is to be the low-cost producer. Be great at **lean**, continuous improvement and process reengineering. Being part of the low-cost play in an oil company with a portfolio of plays is a good place to be right now.

Brush up on networking: Warm up the external oil and gas network before it's needed. If you don't have an external network, it's probably time to start building one. Too many industry professionals tend to concentrate on building their large internal company networks, only to see them vanish when the layoffs start.

KEY TAKEAWAYS

Technology people maintain that God created the world in just six days because He didn't have an installed base. Digital is proving to be a challenging people-change initiative.

1. Oil and gas struggles structurally with change. The nature of the industry has resulted in a sector culture that is risk-averse and, as a result, change-averse.

2. Individual companies face their own specific change challenges, and the nuances of their culture will need to be addressed for change to be successful.

3. The well-intentioned but problematic hedgehogs are commonplace throughout the industry and must be managed carefully to temper their ability to thwart change.

4. Smart machines and algorithms hold tremendous promise for the industry, and building trust in machines accelerates their acceptance and deployment.

5. Leadership needs to be aligned with and committed to digital, which requires a base level of knowledge and skill in digital to be effective.

6. People should always come first. The workforce is experiencing a very turbulent work world and must be supported through the change. There isn't a standby digital-ready talent pool that awaits the opportunity to join the industry.

Notes

1   Aaron Holmes, "533 Million Facebook Users' Phone Numbers and Personal Data Have Been Leaked Online," *Business Insider*, April 3, 2021, businessinsider.com/stolen-data-of-533-million-facebook-users-leaked-online-2021-4.

2   Jose Pagliery, "The Inside Story of the Biggest Hack in History," CNN Business, August 5, 2015, money.cnn.com/2015/08/05/technology/aramco-hack/index.html.

3   Sam Shaed, "Google Is Using Its Highly Intelligent Computer Brain to Slash Its Enormous Electricity Bill," *Business Insider*, July 20, 2016, businessinsider.com/google-is-using-deepminds-ai-to-slash-its-enormous-electricity-bill-2016-7.

4   Bernard Marr, "How BP Uses Big Data and AI to Transform the Oil and Gas Sector," Bernard Marr & Co, 2020, web.archive.org/web/20210408210811/bernardmarr.com/default.asp?contentID=1378.

5   MacGregor, "Husky Energy Deploys Ambyint's AI Technology."

6   Jaremko, "Canadian Natural Planning Test of Autonomous Oilsands Heavy Haulers."

7   "Blockchain: The Key Technology of Tomorrow," Newsroom: The Media Portal by Porsche, November 1, 2019, newsroom.porsche.com/en/company/porsche-blockchain-technology-opportunities-digitization-16800.html.

8   Shimon Y. Nof, ed., *Handbook of Industrial Robotics*, 2nd ed. (New York: John Wiley & Sons, 1999).

9   In brief, will a self-driving car be forced to kill its passengers in order to spare a bystander in the event of an accident? Judith Jarvis Thomson, "The Trolley Problem," *The Yale Law Journal* 94, no. 6 (May 1985): 1395, doi.org/10.2307/796133.

10 "Tech Firms Ramp up Efforts to Woo the Energy Industry," *The Economist*, March 16, 2019, economist.com/business/2019/03/16/tech-firms-ramp-up-efforts-to-woo-the-energy-industry.

11 Peter Jackson, *The Lord of the Rings: Return of the King*, Blue-Ray, Fantasy (New Line Cinema, 2003).

12 Joe Samson, "The Impact of Oil Prices on Calgary's Home Values," JoeSamson.com, November 18, 2014, joesamson.com/blog/the-impact-of-oil-prices-on-calgarys-home-values.

13 "Automation," Volvo Group, accessed May 20, 2021, volvogroup.com/en/future-of-transportation/innovation/automation.html.

14 Sara Reardon, "Rise of Robot Radiologists," *Nature* 576, no. 7787 (December 18, 2019): S54–58, doi.org/10.1038/d41586-019-03847-z.

15 See, for example, Osperity, which offers "AI driven intelligent visual monitoring and alerting for Industrial operations"; osperity.com.

16 BOE Report Staff, "Oil and Gas Leaders Collaborate Using Blockchain Technology to Help Cut Costs," BOE Report, February 14, 2018, boereport.com/2018/02/14/oil-and-gas-leaders-collaborate-using-blockchain-technology-to-help-cut-costs.

17 W. Chan Kim and Renée Mauborgne, *Blue Ocean Strategy: How to Create Uncontested Market Space and Make the Competition Irrelevant* (Boston: Harvard Business School Press, 2005).

5

CASE STUDIES IN DIGITAL ADOPTION

"Before we embarked on the digital journey, we did a lot of analysis. What we didn't anticipate is how 'digital' will scale us up into a much larger market opportunity."

JAMES RAKIEVICH, CEO, McCoy Global

I N RESEARCHING this book, I found many companies keen to share aspects of their digital adoption efforts. While there are important similarities between them, no two companies were exactly alike in their approaches, which underscores that there is no silver bullet. Digital is hard work.

The case studies cover a good mix of the value chain, and the companies have operations across most regions around the world. They are

- an international equipment supplier, McCoy Global, who supplies the upstream sector;

- three North American upstream companies: Imperial Oil, Jupiter Resources, and NAL Resources;

- a North American midstream startup, NorthRiver Midstream, that was carved out of Enbridge;

- a European downstream products company, VARO Energy;

- an integrated international producer, Repsol; and,

- two international services companies, Wood Group and Worley.

Each case study presents four or five key strategies that the company has used to fuel its digital transformation. The chapter includes a summary of tactics that companies may wish to consider as part of their own strategies. The guide that I followed in conducting the interviews is included as an appendix. You can leverage this guide to gauge your organization's progress on digital acceleration.

CASE STUDY COMPANIES

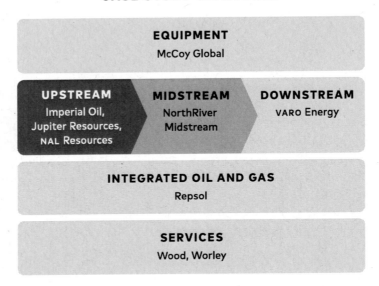

EQUIPMENT
McCoy Global

UPSTREAM
Imperial Oil,
Jupiter Resources,
NAL Resources

MIDSTREAM
NorthRiver
Midstream

DOWNSTREAM
VARO Energy

INTEGRATED OIL AND GAS
Repsol

SERVICES
Wood, Worley

Equipment Suppliers

Case Study 1: McCoy Global

Innovation in oil and gas begins with the equipment industry. Companies in this area of the industry are under constant pressure to innovate as they compete for new business. The upstream sector, in particular, is very dynamic because equipment wears out quickly under the extreme conditions typical of field work. More importantly, the upstream sector has experienced significant labor issues stemming from a wave of retiring senior employees, as well as an inability to attract and retain new talent given the sector's overall volatility. Equipment suppliers to the upstream are heavily impacted by the commodity cycle—equipment demand goes down alongside falling oil and gas prices. Add in the opportunities that are afforded by digital, and equipment suppliers find themselves on the frontline of change. As Jim Rakievich states, "It used to be so easy."

From its roots as a blacksmith business over a hundred years ago, McCoy Global gained industrial manufacturing know-how first in the automotive industry making trailers and springs for the transportation sector. Fifty years ago, the company found itself machining parts for Canada's growing oil sector.

Today the company is a leading supplier of tubular casing makeup gear for the onshore and offshore upstream industry. Tubular casing products are heavy seamless steel tubes, usually cylindrical, that are placed in the wellbore to "case" the well and then cemented in place. Segments of forty-foot-long tubular are coupled in a string with carefully controlled levels of torque or force to hold the segments together during casing running operations. The threaded casing connections must be airtight and become a critical component to the wellbore integrity throughout the production life of the well. McCoy is a global leader in providing the tools, control systems, sensors, and calibration equipment in this area of operations. Their story is illustrative of the challenges facing the tools and equipment suppliers to the industry.

In 2018, with commodity prices in North America still depressed under the lingering effects of the 2015 price collapse, the company purchased a sensor business whose expertise was collecting data in harsh environments in real time for heavy industry, a prescient move as digital innovation in oil and gas was starting to ramp up.[1] McCoy then embarked on a repositioning of its products and services to take greater advantage of its sensor capabilities.[2] To do so, it engaged in these key practices.

DEFINE A ROADMAP

Where larger companies may afford to undertake multiple small pilots and trials to see what works, small companies do not have the luxury to misspend their scarce capital, particularly where product renewal is involved. It takes time to conduct product research, carry out lab trials, develop prototypes, conduct field trials with willing customers, and transition manufacturing, across multiple products and services, all while keeping a legacy business running.

To avoid miscalculation, wasted R&D, and runaway costs, McCoy has invested time and energy to develop a year-by-year expenditure plan to reposition its products and services. Its vision is to digitally transform the business to improve efficiency, enhance safety, and deliver a more environmentally acceptable result. The plan extends out five years because of the product-development timelines. The roadmap is a board-level document, reflective of the investment of shareholder capital in the redesign of products and services, and has the personal oversight of the CEO, supported by the CFO (capital resourcing, cost control, financial modeling, and reporting), and by the VP leading sales and technology (project execution and market alignment).

To underscore the seriousness of the changes sought, the CEO holds weekly meetings with the management team to review progress, clear roadblocks, and solve problems. There is no substitute for a high-caliber, supportive, and aligned leadership team when it comes to executing strategy.

RECOGNIZE THE VALUE OF DATA

It's a meme now to state that "data is the new oil," but for tools companies whose gear has traditionally generated usage data, technology limitations have made it so very few were able to capture that data directly. Purchase contracts often did not distinguish between the tool being purchased and the data that the tool generated. In most instances, the data was assumed to belong to the customer who purchased the tool. They alone had the telecom network at the work site and the computer capacity to store the data generated. New technology upends that orthodoxy.

For many years, McCoy's tools have generated a stream of operating data about the tools and their usage. That data was not considered to have much value except to the customer, and was not collected or analyzed by McCoy directly. That data is now recognized as having significant value, both technically (second by second, recording how the tool is performing), and commercially (job to job, keeping track of where the tool is being used). The data is now considered fundamental to the company value proposition.

Building up critical data capabilities (capture, storage, analysis, disposal) warrants its own component on the roadmap.

IN MID-2021, I was helping a tools manufacturer prepare marketing copy to support their product positioning along environmental sustainability grounds. The tools in question were strictly mechanical in nature.

We organized a briefing with a customer who had carefully studied and measured the very positive environmental impacts of the tools on their pilot site. The customer was very pleased with the tool, but concluded the briefing with a dire warning: "Unless the tool is somehow digitalized so that it self-monitors, it is doomed."

That message did not make it into the media copy.

UNLOCK NEW BUSINESS MODELS

Even small companies can discover and unlock new business models through digital. The tools that large companies use to enable collaborative remote teams (Slack, **Yammer**) are also fully available to small businesses. For example, because of COVID-19, the company has learned that it no longer needs to tie new product development to specific locations where the company maintains a physical presence. The addition of digital smarts to its products can actually be accomplished largely virtually, and from anywhere.

The two big change drivers for well drilling companies are talent and capital, because both are in short supply. A very significant number of experienced drilling talent has permanently left the industry, leaving drillers woefully exposed to drilling cost overage because of inexperienced crews, and a high-cost burden to train up new recruits. The economics of automation alone are very compelling. The capital shortage creates an incentive for drilling companies to consider rental or subscription business models.

Smarter tubular running tools, souped up with sensors and simplified user interfaces, help relieve the talent burden by making the tools not only safer but easier to learn and use, and reduce the need for highly experienced senior rig floor hands to supervise the work. Coupled with cloud computing and better telecom capabilities, McCoy has unlocked a remote tool-use monitoring service, which provides a different revenue stream from the traditional product sales model.

The future business model may feature elements such as

- much better asset tracking that enables a commercial structure based not on purchasing of tools but usage of tools;

- a subscription to the data generated by the tools;

- the removal of people from the "red zone" on the rig to significantly impact safety, training, and retention;

- the opportunity for creative gain sharing with customers because of higher work quality based on better tool use;

- freely distributed tools that do away entirely with the customer capital expenditure; and

- rapid scale-up of the business because of its digital dimension.

Getting the market to adopt these innovations takes more time than you might think. "Initial market adoption is hard work. You can't put enough money into it to make it happen fast," observes the CEO.

ENGAGE THE ORGANIZATION

Small companies perhaps have an easier time communicating their goals and agenda than larger firms, but they must still invest time and energy in that task. After all, many employees and board members have invested personally and emotionally in the success of the legacy business. It is only the startup that does not have a historical business to acknowledge, honor, and defend. Indeed, the past cannot be fully abandoned as long as customers are using older equipment.

McCoy noted that the board and investors need regular reassurance that the company is on the right track. Without positive messages and a sense of excitement about the future, small companies may struggle to attract talent able to help with the transition or capital interested in financing the business.

A few key tactics stand out:

- a single common message shared internally with its employees and board;

- a tuned narrative for the external customer audience;

- a biannual open mic session involving the CEO and other executives in small groups of eight employees at a time (small groups promote more dialogue);

- regular briefings on the state of the industry, the company, and its progress;

- celebrations on achieving critical milestones on the roadmap;

- lunch and learn sessions on key topics of interest, such as cloud computing;

- strong involvement from the head of people and culture in the transformation program; and

- tight working relationships between the mechanical teams who build the gear and the sensor teams that give it smarts.

The company measures its success by the steady demand pull for innovation and digital changes from the field, and from the interest from customers seeking briefings and updates on new product development.

Many companies conclude that these transitions are of little interest to customers, but in the case of new digitally enabled product development, the interest of the customer is critical. First, the customer has to want to make the business more efficient, with many efficiency opportunities in oil and gas seen as too risky. Customers need to be sufficiently motivated to give the new automation solutions a trial. Second, the value of the new data-centric business model may require a different sales approach, away from a merchant of metal goods to a merchant of gigabytes.

Upstream Companies

The person on the street, when asked about oil and gas, conjures up an image of an upstream company, featuring an active drill rig with mud-splattered rig workers, or a slowly bobbing lonely **beam pump** on the prairies. In reality, exploring for resources and subsequently producing those resources involves a complex supply chain of specialized skills, equipment, and services to safely build the well, build the surface facilities to handle the resource, and run both subsurface and surface operations economically. The supply chains vary dramatically depending on the resource—oil, gas, offshore (scale, oceanic), onshore (trucking, **frac spread**s), or oil mining (shovels, dump trucks).

The upstream industry has always been a prolific user of computer technology. It practically invented big data—geologic data models are enormous, and the computing power required to process the data drove the early computer industry. The immense sums involved are equally staggering, and unproductive capital on the balance sheet awaiting attention pushes the industry forward. Digital is now part of the solution basis.

Case Study 2: Imperial Oil

My first job out of university in 1984 was with Imperial Oil, at the time based in Toronto. Rumor had it that the company owned a fine collection of Canadian art, on the top floor of the St. Clair Avenue head office where I walked to work. One Sunday, out for a stroll with my wife, I decided to investigate, and up the elevator we went, on my very first trip to the top-level executive floor. The rumors were true—along the walls were paintings spanning two centuries by Canada's great artists. Within minutes a security guard appeared behind us with a polite "What are you doing?" "Admiring the art," I responded. "Take your time," he said, which I interpreted as "it's time for you to leave." I remember my heart beating a little faster as I realized we were probably trespassing.

One of Canada's largest oil companies, Imperial Oil has long been seen as the leader of the sector, with interests spanning virtually all facets of Canada's diverse energy mix. It has operations in conventional oil and gas in the Western Canadian Sedimentary Basin, oil sands mining in the Athabasca Basin, thermal operations in Cold Lake, as well as refining, wholesale, trading, and retailing. It is the diversity of its upstream interests that distinguishes Imperial Oil from its domestic peers.

The upstream digital team is distinct among Exxon's broader digital efforts because of its leadership. Imperial's CEO initially recognized the need to accelerate digitalization and requested regular updates on progress. The sponsor is the VP of upstream, and has a steering group including the VP, upstream technical manager, IT manager, production unit managers from the assets, as well as planning and execution leaders. A dedicated "fusion" team has garnered the most traction and success among similar teams globally. Heather Wilcott describes the team as "consisting of upstream team members, data scientists, software execution, architects, and data lake team, all co-located. It's unique." Its stand-out practices are its narrative, its focus on data, its perspectives on talent, and its emphasis on change champions.

ADOPT A BALANCED NARRATIVE

The upstream digital team has adopted a balanced narrative to communicate the digital agenda. Openly acknowledging the brutal truths about the industry (the need to decarbonize, the requirement to improve sector competitiveness, the imperative to retain a social license to operate), the team highlights the use of digital tools to maximize its producing assets (its oil sands mine, its heavy oil unit, and other production assets). Through digital, its people can be freed from the mundane daily manipulation of Excel spreadsheets to focus on the assets, solve problems, optimize and improve complex processes, innovate, improve reliability, lower costs, reduce waste, and boost profits.

Rather than explicitly stating a digital strategy, the company aims to have an asset strategy that is enabled by technology, including digital. By embedding digital within the business model and goal set, the team democratizes the adoption of digital solutions and creates the context for managerial accountability for success. The consequence is that the different assets may vary in their digital directions and choices, suggesting an ongoing need for some corporate steering so that the level of digital diversity is moderated.

THE OIL sands mines face a critical bottleneck asset—the mine shovel. If the shovel fails, the entire ore flow halts. A business goal to optimize shovel reliability opens up many candidate solutions, including better shovel materials, more spare parts, and greater surge inventories. Teams are now encouraged to use data science and analytics to better understand failures and to inform operational changes. To help employees keep a clear eye on the business, managers do not have a digital shovel strategy, but rather reliability goals with technology enablers in their tactical plans.

GET BEYOND DATA

Through the digital learning journey, Imperial discovered quickly that leveraging data to prove a point and improve the business was the key to unlocking manager interest in digital. Once managers can see how high-quality data speedily analyzed with modern tools enables faster decision-making about profitable business change, they quickly become advocates for more of the same.

The company's attitude towards data is now completely transformed. Investments in improving data foundations are a major spend category, to bring about the integration and contextualization of diverse datasets, and to use those datasets to create value. CAPEX now flows to such areas as transport infrastructure (the technologies for moving data to and from operations and the edge), cloud adoption (only cloud systems can handle the volume), data lakes (the immense datasets), data governance (decision-making about things like taxonomy, standards, and ownership), and data expertise.

EMBRACE A BROADER VIEW OF TALENT

Early in its adoption program, it became clear that the company did not have the full range and depth of skills it would need. To kick-start their digital efforts, Imperial leveraged a team of consultants to frame the pathway to follow, but it has since advanced into building up in-house digital capabilities that work closely with its outsourced teams based in India. This focus on building capacity explicitly acknowledges that an understanding of how the current state works (technology, process, data, people) is crucial to improving it and enabling it with digital tools.

In the adoption of digital tools in an industrial setting, think of the talent model like an iceberg. Plainly visible above the water line are the high-profile data scientist jobs, a role now in every industry, not just oil and gas. But what is not so visible, below the waterline, and in far greater numbers, are the roles to provide the rich and robust data volumes that the data scientists need. Imperial Oil notes that these jobs are many and varied, important, and in demand, particularly across a sprawling upstream portfolio. They include

- telecommunications engineers who assure **fiber** and wireless connectivity;

- instrument specialists who install and maintain the sensors;

- control system engineers who keep operations running and enable access to data;

- ERP system professionals who run the large commercial systems;

- data engineers who solve data challenges;

- system architects who plan out applications and integrations; and

- software developers who build and contextualize datasets and help enable data access across disparate systems.

Beyond boosting its internal capacity, Imperial's long-term goal is to make digital a part of everyone's job, upskilling workers as citizen analysts—enabling them to connect to datasets and build their own **visualization**s and analyses to support their everyday decision-making. This avoids creating a capacity constraint by housing all data work in a central data team. It takes training, different tools, and strong support. Finally, the talent pipeline for the kinds of new data skills the company needs (data scientists, data engineers, data architects) is being enhanced.

DEVELOP DIGITAL LEADERS

Imperial recognizes that digital success requires a clear business case, the efforts of a range of business professionals and subject matter experts to help identify and solve the problem, a technology team that can stand up a solution, and the ability for all to work together quickly. And managers and supervisors have to want to change.

Like all industrial enterprises, Imperial Oil's culture is rightfully skeptical of digital. Its managers want proof, not promises. Its leaders are widely spread out, are full-time busy running their day-to-day, likely on shifts and in hard-to-reach places, and their assets are all different. Their teams are at capacity. They typically have little budgetary room to experiment. Their reward systems motivate them to value stability, reliability, and safety, and to adhere to their agreed processes. Change, including digital change, often requires significant effort from already taxed resources and may not take priority over the challenging day-to-day operational activities that must be managed. There is simply no one thing that they can all do, nor is there one PowerPoint presentation they can all receive, that magically alters this reality.

To accelerate the digital journey, Imperial needs to convert as many of its digital doubters as it can into digital advocates. "Our CEO is always asking, 'What do you need to go faster?'" says Heather Wilcott.

The company takes a multifaceted approach to this cultural shift. The digital team is full-time so that it has focus and no distractions. It is led from the top to reinforce the importance of digital change. The team includes business professionals as well as digital and data experts, and opportunity managers who help lead the idea-to-solution process. The team patiently works with managers and leaders to help move them from uninformed to a digital trial. Training dollars are set aside to teach the concepts. Reasonable risk-taking is encouraged. The successes get a lot of attention on the internal intranet, the Yammer channel, and the regular briefings.

Case Study 3: Jupiter Resources

Jupiter Resources was a natural gas production company principally owned by private equity sponsor Apollo Global Management of New York, with operations in the resource rich Grande Cache area of Alberta, and the home office based in Calgary. Established in 2014 with the purchase of the Bighorn gas assets from Denver-based Ovintiv, Jupiter peaked its production at 80,000 barrels of oil equivalent per day (**boe**/day), and was eventually sold to Tourmaline Oil in late 2020.[3]

Named after the largest gas planet in our solar system, Jupiter aimed to grow into a gas giant itself, an inspiring goal in a world keen to move off coal for climate reasons. Commodity markets, however, are ruthless, and while Henry Hub gas prices started in 2014 at an optimistic $4.50, Canadian markets temporarily dipped to zero during the following years, before starting to rise again in late 2020.[4] This underscores how critical it is for oil and gas businesses to run lean.

Jupiter stood out for its unheard-of staffing model of fifty home office staff and fifty workers in the field. This yielded a production rate of about eight hundred barrels per employee. In comparison, Ovintiv's production in 2019 was 573,000 boe/day with an employee headcount of 1,916, or three hundred barrels per employee.[5] Patrick Elliott says, "Digital helps you put more engineering time back into the hands of the engineers."

WHILE I was on a brief consulting project with Jupiter, the EVP sponsoring my work extended to me an invitation to a "rough" session. In my career I had organized many meetings best described as rough, as in "that was a rough meeting," but generally I only judged meetings as rough after the fact. It turned out that "rough" was actually "RUFF," which stands for Report the Ugly and Find the Fix. It was a problem-solving discussion that pulled people from across the company to help find solutions in a constructive and supportive environment, avoiding the name, blame, and shame approach. Either way, how could I say no?

Immediately on launch, Jupiter's management team set out to create a different kind of energy producer. The management team had considerable prior experience with other startup and junior oil producers, and it wanted to avoid creating yet another undifferentiated industry player in a crowded gas producer field. These are the key practices it followed.

USE DIGITAL TO SUPPORT AND REFLECT THE CULTURE

As a startup gas producer, Jupiter was in the enviable position to articulate the culture it wanted to build and to take a series of explicit decisions that served to create that culture, including very explicit digital foundations.

The cultural target was an open, transparent, and collaborative operational environment, so all employees could contribute their talents fully to achieving the company's goals. Talented people with access to all the information they need to do their jobs create deeper insights, surface more opportunities, and solve problems faster, using fun sessions like the RUFF meeting. Swift and available access to operational information was critical to achieving this cultural expression.

Gravity actually pulls traditional upstream oil and gas companies in the opposite direction. Application systems choices can trap data inside applications, and high per-seat pricing models block wide access to that data. Friendly internal competition structures that are intended to surface the best opportunities for capital allocation can sometimes cause employees to not share information openly. Failures in the business can be viewed as signs of poor performance, leading employees to not share problems seeking solutions.

The result is a kind of information asymmetry, where most employees are subtly ill-equipped to do their jobs, some employees have information privilege, others have application privilege, and much time is wasted simply trying to reconcile why two sets of numbers don't match. Problem-solving is slower, as professionals first have to work on the data rather than the problem. Much productive energy is lost in carrying out such mundane tasks as preparing quarterly board reports.

In Jupiter's vision, business data became the critical corporate asset. Employees needed to be taught their new relationship with data, that it was owned by the corporation, not the department from whose system the data originated, and not an employee. Data was to be organized to be freely accessible, not to be hoarded. A single version of the truth was the goal, and having multiple independent and slightly different versions of the same data was anathema. Data was to flow freely and swiftly, unimpeded by gatekeepers, arbitrary system restrictions, and technology choices.

Digital innovations that supported and enabled this cultural target included:

- cloud-enabled apps and centralized data storage and access, which allowed employees access to data to do their work;

- mobile access to free up decision-making;

- electronic data capture from remote sensors anywhere they are located; and

- common toolsets between field and head office to facilitate collaboration.

Many companies use common digital tools but fail to achieve a business culture of transparency in their operations. Business leaders should not assume that tools alone will change the business for the better.

EMBRACE MODERN INFORMATION MANAGEMENT

With data playing a key role in achieving the desired culture of business transparency, the choice of organizing methods for data became an important executive discussion topic. The prevailing model of information management at the time is still very much familiar today— email threads, web browsers, hierarchical server file structures with odd file names and size limits, file extensions, hard-to-use metadata, uncontrolled file replication, awkward access privileges, no internal search engine, separate viewers for images, PDFs, spreadsheets, recordings, and videos. This approach contributes to bloated data holdings, unsearchable records, and uncertainty of provenance.

The leadership became convinced that continued reliance on the traditional tools and methods of the industry to create a fundamentally different culture was likely to fail. The norm actually appeared to be counter to the culture they wanted to create.

An alternative solution to the web browser model, called Share-Point by Microsoft, had been evolving for many years, and was released in 2013 with many of the features that seemed to align with the cultural target—ample social collaboration, **wiki**s, shared calendars, document libraries, shared task lists, surveys, and mobile access. Adoption would take many months as the employees and executives became accustomed to the different experience imposed by Share-Point, but once adopted, there was no going back. Eventually, Jupiter moved to the full cloud versions of SharePoint and the Microsoft productivity suite.

Microsoft Teams, which was launched in 2017, unlocked real-time collaboration between the two offices, with the field office rapidly deploying Teams separately to administer activity between the field office and the Calgary office teams. This proved to be a pivotal solution for survival during the pandemic.

Swift and free data access only partly fulfills the cultural target. The company also needed to dramatically boost the analytics capability to take advantage of all of the data that was accessible. Early on, a small data analysis pilot using **TIBCO Spotfire** software proved convincingly that analytics skills were likely within reach of most staff, and in time the company created an army of citizen data analysts. Over 70 percent of employees eventually had licenses for daily use of the analytics software. It was commonplace that each operational group had a Spotfire expert who could integrate datasets across a number of different disciplines and could create insights and actions that improved efficiency and understanding.

ENGAGE AND SUPPORT THE FIELD

It is common in oil and gas for deployment of innovation to struggle in a field setting. Field employees often work shifts, which makes training costly. Field assets may be some miles distant from the field office, creating logistical challenges. And with so many workers in the field, there often isn't enough office space to assemble all the workers for meetings anyway. Operational problems take priority over all else and pull employees away to secure the business and protect the environment, leading to uneven training outcomes. Travel to the home office for training is costly and impractical given the numbers. Tools that work well in an office context with big screens and keyboards may not meet the needs of the field.

Jupiter's cultural goals aimed to create a coherent organization that included the field. This was accomplished through a number of tactics:

Use the same tool sets: The same collaboration tools (SharePoint, Teams) and analytics tools (Spotfire) were standard in both settings.

Incorporate mobility wherever possible: Mobile versions of key solutions and software were prioritized.

Deploy robust telecom access: Telecom capabilities were upgraded to provide a clean and even level of service, and to enable data capture from remote IoT devices. Effectively, real-time data provided the insights that gave the field more meaningful actions to take.

Give the field responsibility for key data: The field was assigned responsibility to manage critical data originating in the field directly in shared corporate systems.

Include the field in key decisions: Any governance structure put in place to deal with digital and systems topics included influential field employees that drove greater field and home office integration.

Virtualize the data: Moving enterprise data to the cloud allowed the field to see and work with the same data as the home office team and enabled a superior level of collaboration across the company.

Digital helps remove the time and location barriers that create the divide between the head office and the field, resulting in a better working relationship.

NURTURE THE ECOSYSTEM

As a startup, Jupiter's management team was empathetic towards other new businesses and felt a level of responsibility to explicitly nurture the ecosystem of small independent startups that were able to offer solutions of value. Management created the expectation of employees to seek out solutions to their identified problems, under the assumption that collaboration with any emerging market solutions was a superior approach.

The company worked with a variety of hard technology companies (physical devices and equipment) as well as digital technology to trial new ways of working and explore innovations. As with any portfolio, some conversations did not go far, but a substantial number of companies wound up in a commercial engagement with Jupiter, with a few even able to claim Jupiter as their first customer. Some promising ventures achieved financing because of the strength of their relationship with Jupiter.

Jupiter approached their innovation agenda much like a venture investor would. For example, they nurtured an investment opportunity funnel that gave them exposure to a range of startups spanning seed level to commercial viability. To balance the portfolio, they invested in technical cooperation, pilot projects, and full commercial engagement. Progress tracking through the commercialization stages helped demonstrate how the investment in technologies created value for the organization. Finally, Jupiter co-authored white papers with their partner technology companies about the positive outcomes being realized, which served to lend credibility to these new technology businesses.

It's likely Jupiter's relatively small size made such engagement easier to manage than might be the case were Jupiter much larger, but the move underscores that even smaller oil and gas producers should engage with the digital ecosystem as a priority.

Case Study 4: NAL Resources

NAL Resources was an oil production company based in Calgary, Alberta, and owned by Manulife Financial, a Toronto-based financial services company. NAL started life in 1990 and survived the myriad gyrations of the industry throughout that time. In late 2020, the company was purchased by another producer, and the brand has since vanished.

For a time, NAL served as a digital role model for many similar-sized oil and gas companies in Canada and the US. Digitally enabled change emerged as a solution candidate at a three-day offsite the company held in mid-2015, where the executive team concluded that oil prices were likely to stay low for much longer. The 2014–2015 oil price collapse forced the upstream industry to retrench. NAL dusted off the classic industry playbook and sketched out a lengthy survival strategy to lower costs, boost revenues, and create more opportunities. They eventually adopted a series of new behaviors, operations practices, and digital innovations that proved to be considerably more valuable than originally thought and inspired other organizations to trial some of their ideas.

THE CFO for NAL Resources, who was sponsoring NAL's digital program, agreed in October 2018 to appear on my podcast to discuss their transformation agenda. This was unusual in two respects. First, the buyers of digital innovation in oil and gas rarely grant interviews because of the tendency to view their internal efforts as competitive. Second, the CFO is normally the executive most concerned about company confidentiality and charged with enforcing company rules about disclosure. In the interview, the CFO made it clear that the most important digitally enabled changes required whole-of-industry support and not just the efforts of one or two pioneers. It was suddenly an imperative to share their story.

It took a couple of years before NAL's digital program started to show real progress, a warning for other oil companies not to halt energies prematurely. Indeed, NAL initially bet on finding a transformative acquisition that would dramatically reshape the portfolio of assets, but when it became apparent that such an acquisition may not transpire, efforts to fix the existing business stepped up a gear. In time, NAL embraced a wide range of moves to accelerate its digital shift, with the most impactful among the following tactics.

YOKE DIGITAL CHANGE TO BUSINESS CHANGE

For many an oil and gas company, the most widely pursued path to higher performance, a transformative acquisition, can provide hope that the business will survive without a lot of change to process, people, and technology. After such a deal, mature high-cost assets can be sold off once the new low-cost assets are in place. NAL acquired several properties, gaining some general and administration (**G&A**) savings, but nothing transformative. The company then had no option but to accelerate the more painful business changes necessary to restore the existing portfolio and business architecture to meet its performance goals.

Recognizing that the cost structure of the business was a direct result of choices that the company had made, the executives resolved to make production choices to drive costs in the right direction by

- focusing exploration on key assets where geographic concentration and higher working interest were achievable;

- boosting per-well productivity by focusing on opportunities with higher production potential;

- lowering overall CAPEX by drilling fewer but longer wells;

- reducing surface CAPEX by spacing wells closer together and sharing surface infrastructure;

- minimizing surface disturbance by reducing the well pad footprint; and

- improving carbon resilience by concentrating emissions-prone infrastructure and retrofitting existing assets with low- or zero-emissions features.

As Cory Bergh frames it, "With your digital investments, target decision-making that drives cost and productivity, and not the cost itself."

Meanwhile, NAL's initial digital change efforts were unremarkable in scope and impact, but over time it became apparent to the executives that digital innovations were both economic in their own right and multiplied the impacts of its new business choices. The CEO was then able to confidently state a compelling new vision for the organization. Coined as the "operator of the future," the vision was achieved by embedding digital technologies deeply into the wells of the future, helping them perform like premium assets to

- reduce OPEX by managing all oil production entirely remotely;

- reduce OPEX and risk by visiting wells only when prescribed by regulation or when absolutely necessary; and

- maximize revenues by achieving 100 percent of theoretical well production 100 percent of the time.

To reduce well visits, the company needed to shift from reacting to well events to proactively managing its wells. NAL first moved all wells onto a consistent SCADA platform, which yielded consistent data about well operating state. This data was fed into a PI Systems historian and data repository, and enabled an algorithm based on IBM Watson to predict well events with accuracy. Operator tasks followed the priorities set by Watson, and directly led to around 15 percent improvement in value add in a key location.

The new operations analytics team soon discovered that they were suffering from the traveling salesman problem. The daily task list was no longer based on the minimum driving time established by the intuition and experience of the operator, but on the highest-value but non-intuitive activities as estimated by the algorithm. Driving time then increased, adding to a major safety risk and contributing to emissions. Turning again to IBM Watson, and accessing private road data from government sources, the operators pooled their collective expertise and helped develop an optimal routing algorithm that in a limited implementation saved the operators another 15 percent in costs.

Since operations in oil and gas are responsible for the bulk of operating expenses and capital spend, real change starts only once operations captures meaningful benefits from digital innovation.

ACCEPT THE LIMITATIONS OF PEOPLE

Oil companies generally hire people to work in the business, to follow the procedures laid down, to abide by the regulatory rules, to comply with instructions, and to solve the many daily upsets in a way that brings the business back to its steady state. Rule-breakers can get companies and themselves into trouble. It is widely held that any failures in such a rigorous business must be due to poor performance and represent a waste of company resources. Daily meetings such as the **Safety Moment** reinforce adherence to procedure to keep people, the community, and the environment safe.

This culture gets in the way when the business needs to reinvent itself. Companies find themselves without a sufficient number, if any, of the skills needed to run experiments, trial new techniques, develop new solutions, accept errors, and take risks.

NAL elected to confront this challenge as part of its digital change program with these guiding principles:

- Understand that the new ways of working will be fundamentally at odds with the original employment contract.

- Acknowledge that many employees will not be inspired by the prospect of change, and not motivated to work on initiatives that upset their legacy work practices.

- Recognize that employees lack the skills to conceptualize new ways of working and do not have time to gain those skills. Design a program to address this issue.

- Gracefully accept that some employees will be incapable of working in different ways and help them find success elsewhere.

The digital change leaders actively sought out those individuals who came across as believers in the change agenda. The company brought in digital influencers (community thought leaders, vendors) to present to the employees on important digital topics, like agile ways of working, in the hopes of inspiring some interest.

One proxy indicator for individuals who "got it" surfaced over the debate about applications versus data. Much of NAL's oil and gas data was trapped in software application silos, with different definitions, interpretations, rules, and acceptable values. NAL learned that 50 percent of the common data the company had in its various systems was materially different between systems. Data mature individuals recognized that it was the data, and not the application, that held the value.

CREATE A PROGRAM

NAL's digital program began to move in earnest when the company rethought its approach to getting initiatives going. The digital leader sought and was granted a budget, but how to spend it given the talent limitations remained a big question. Hiring was not an option under the conditions. Waiting for good ideas to bubble up was too slow given their situation. "If a technology company owned an oil company, how would they do it differently?" asked Cory Bergh.

The resulting program was simple—document a change idea in a consistent fashion, fund it, get it approved, and track its results. The company created a strategy desk whose role was to serve as a concentration point for all strategies and change initiatives, including digital. The desk crafted a simple funding request form that streamlined the approval cycle. Kept simple, the description of the initiative was a maximum of thirty words. The desk provided advice and guidance to anyone pursuing a digital change. Finally, the desk reported results every quarter to match the regular internal management and board meeting structure.

One powerful change was to reverse the approval tree. Rather than the slow, traditional approach (get your manager's approval, then get their manager's approval, and upward eight levels), the digital leader, a vice president level, gave immediate approval to a thirty-word summary request for a $100,000 budget allotment for G&A changes, or up to $250,000 for capital. With this in hand, the managers below the VP could quickly accept the risk, and the citizen entrepreneur could get moving.

The results of the program became so impactful that the funding doubled in the second year, and doubled again in the third. The CEO confidently presented to public audiences that digital programs were producing benefits that frequently exceeded the best available capital investment in new wells and facilities.

Digital results began to meaningfully impact the ability of the company to achieve the targets set out for the employee bonus program (called the Short-Term Incentive Program or STIP), and employees could see the effects the digital efforts had on their compensation.

RUN LOTS OF TRIALS

By "lots," I mean that NAL was running forty to fifty separate digital trials, ranging from zero cost, other than employee time, to hundreds of thousands in capital investment in digital for the field. The goal was always to get a digital change into production, not necessarily to scale where it was deployed all across the company. Solving for scale would come later.

For example, NAL worked with GuildOne, a royalty advisor, to develop a smart contract to settle a royalty owing. Many of NAL's wells were structured with multiple venture partners, and every month, production accounting needed to pull the contracts, review the terms, apply the production volumes and costs, and pay the royalties.

Getting all of its counter parties to work on this initiative at the same time was highly unlikely, given the volume of wells, the levels of interest in the concept, and the level of outstanding disputed royalties. Instead, the company collaborated with just one smaller royalty firm to prove out the concept and get the smart contract to work correctly. Initially, NAL learned that only 10 percent of its data correctly matched the same data held by the royalty firm. Using a combination of RPA, machine learning, and blockchain technologies, NAL was able to clean up the data and eliminate disputed royalty calculations with its counterparty. Putting the process into production (although not at scale) improved the acceptability of the idea in the industry.

Once proved, RPA technology rapidly spread, and the company eventually had twenty-three separate robots carrying out work, each with a very positive payback. Over five hundred smart contracts were concluded and put into production. A further fifteen fully digital workflows were deployed, including an end-to-end digital source-to-pay process. Some five million digital documents were housed in **data marts** that provided business intelligence to the whole of the company. Eventually the finance function boosted its staff productivity by over 50 percent. Production accountants moved from 125 wells each to over five hundred per month, with no reduction in **KPI** performance.

Over time, the outlines of the back office of the future came into view:

- Parties no longer dispute data (data is consensualized between parties).

- Data is pre-reconciled to eliminate after-the-fact reconciliation effort.

- Work is fully automated, removing people as much as possible.

- Data resides in the cloud so as to enable a future of machine learning AI.

- Employees supervise robots, a suite of electronic workflows, digital documents, and business intelligence reporting.

Midstream Companies

The midstream sector is principally composed of large industrial assets for moving hydrocarbons about, and carrying out various kinds of treatments to the hydrocarbons to increase their value. The list of these kinds of assets is unending—**separation unit**s, gas plants, **straddle plant**s, **compression station**s, pipelines, refineries, rail loading and unloading, and ports. In the main, these assets are driven by decisions that keep the assets running as steadily as possible, at the highest possible utilization rate, at the lowest possible cost, to the required quality level, for an optimal turnover rate.

Case Study 5: NorthRiver Midstream

A carve-out from Enbridge, Canada's largest pipeline company, NorthRiver Midstream (NRM) is owned by Brookfield Infrastructure Partners, a global owner and operator of infrastructure networks that help move energy, water, goods, people, and data around the planet. Enbridge, in a major series of asset sales following its acquisition of Spectra Energy, reached an agreement to sell the NRM assets in mid-2018, with the deal closing in December 2019.[6] NRM was subsequently formed to own and operate nineteen natural gas processing plants and over 2,200 miles (3,500 kilometers) of natural gas **gathering pipeline**s, principally in the **Montney** region of British Columbia.

NRM is a fascinating case study of the potential of using digital innovations to transform mature assets. Some of the gas plants are decades old and would conventionally be viewed as beyond innovation. However, the deal to carve out the gas processing assets excluded any related information processing for the assets, including their SCADA systems. From day one, the company embarked on an agenda to do energy differently, which includes its approach to digital adoption. Its key moves included the following strategies.

CREATE A TRANSFORMATION NARRATIVE

A transaction, such as a divestiture or an acquisition, demands an explanation, which simultaneously creates space for an impactful new narrative about the need for change. NRM exploited their transaction masterfully, with the branding "Energy. Done differently," as the foundation for a conversation with employees about the need to operate the company on a basis at variance with industry norms.

The conversation included the following points:

Customers have more options: In the past, customers signed long-term contracts with midstream players, but the growth of the industry along with modular asset construction gives customers flexibility to use other suppliers or build out their own midstream assets. The costs of the traditional ways of working can create the incentive for customers to explore these alternatives.

Operations needs to simplify: The traditional approach, with each plant effectively operating independently, lacks scale, and is complex to manage. Costs are duplicated throughout the business. Businesses don't sell off their crown jewel assets, and the implicit message from the transaction was the need and opportunity to do better.

Talent is uninspired: It's hard enough to attract the brightest engineering talent to work in energy in the face of digital opportunities, and it's doubly hard to work on sixty-year-old assets frozen in time. In recruiting interviews, young professionals ask specifically what platforms the company is using.

Digital is an imperative: The level of automation and autonomy needs to step up to match an ambition to be world-class. This is an inspiring message for those employees interested in being part of something bigger.

ESG is important: Access to capital is contingent on demonstration of commitment to ESG goals and objectives. The tools of the past are not fit to help understand the emissions footprint, and more importantly, constrain the available solutions. NRM could confidently point to its new suite of the latest world-class tools and technology and articulate how these new ways of working could fundamentally position the company for the future.

GO CLOUD ALL THE WAY

Is it possible to migrate a midstream company completely to third-party cloud infrastructure? Yes, if you have the opportunity.

IMAGINE YOU are the new **SVP** for operations of this sprawling venture. On your first day, you visit the floor of the office where you will be located when the deal closes, and you learn that you have no computers at all, no Wi-Fi, no servers. You learn that you will not have any SCADA systems when the deal closes in a few months. You resort to using your mobile phone as a **hotspot** so you can get work done.

NRM faced an important decision point on its architectural choices when the company was launched—replicate the safe decisions of the past and build out traditional infrastructure (control rooms, server closets, desktop devices, phone handsets) or build for the future.

With its branding statement and the goal of becoming a platform for growth within Brookfield Infrastructure, NRM wanted to demonstrate that it was an asset worthy of expansion. It elected to build out the business with the future in mind, and cloud computing was adopted as the sole platform model. And not just for the commercial systems such as its ERP platform and desktop systems, but for everything, including its SCADA systems.

This cloud strategy enabled NRM to transform the business. At the time of the deal, the gas plants were advantageously located close to each other in the same region, but were all on separate SCADA versions and had different safety programs, separate operations procedures, and separate suppliers providing services.

Moving all the assets to the same cloud-enabled solutions provided the basis to harmonize operational procedures, adopt a common safety program, and gain scale economies with the supply chain.

Importantly, NRM will soon be able to stand up an integrated remote operations center (**IROC**) to supervise all the facilities from a single location. With the IROC using the cloud, operations will have visibility into plant performance from any device, anytime, anywhere. As Jay Billesberger puts it, "I can run gas control from my phone from anywhere there's an internet connection if I wanted."

All of this migration activity took place principally in the pandemic year of 2020; it was a time of great challenge in achieving results, but NRM realized a 20 percent reduction in costs from its new approach. For example, the company deployed a 100 percent invoice automation process that reduced the number of accounts payable clerks by 80 percent. The few remaining staff review invoices for completeness and routing efficiency only.

DO SUPPORT DIFFERENTLY

A hallmark of companies that are leading in the successful adoption of digital innovation is how they choose to organize support for what could be multiple technology domains:

- Information technology (IT): the world of commercial transaction systems, help desks, cyber defense, networks, and servers.

- Operational technology (OT): the world of industrial control and monitoring systems (SCADA, microwave towers) running 24/7.

- Digital technology: the world of sensors, cloud computing, mobile apps, and social media.

NRM very deliberately supports its technology environment in a blended model. There is one SVP responsible for business operations and technology, who reports directly to the chief operating officer

(COO). Within operations and technology, a single director, reporting to the SVP, holds responsibility for all computer technology (information, operational, and digital), which dramatically accelerates decision-making. For example, a request to access the PI Historian for a data mart project in a traditional structure can often require two VPs to clear the path for progress. In NRM's case, the director owns all the platforms and can make faster decisions.

Some of the practices that have made this support model work include adopting selected agile methods to accelerate change. Now the executive team is following some of the same methods, such as a daily fifteen-minute scrum meeting to keep the leadership team aligned on business priorities.

MANAGE THE PEOPLE TRANSITION

The pandemic helped open up the organization to consider alternative job models, such as carrying out work virtually from anywhere. Like many oil and gas businesses with long-life assets in isolated locations, NRM still faced the following significant constraints in its ability to drive change:

Innovation shortfalls: Long service operational employees may lack the skills in innovation necessary to help with business change. Over time, long-life **take-or-pay** contracts eliminate the incentive to innovate, contribute to a closed mindset, and may even foster a culture of hubris.

Lack of enterprise data skills: The traditional model of decentralized independent assets running on their own does not prioritize enterprise data skills.

Small-town constraints: Aggressive job changes could be devastating for workers and their families. In small towns, both spouses may work for the same company, and there may be few alternative employers.

The hard change is actually the data transition. Providing for people impacted by digital change and helping those unable or unprepared for business change are signs of a well-run and ethical business. But boosting data acumen among the hundreds or thousands of employees is the heavy lift.

At NRM, effective data delivery meant all employees needed to become more interested in and concerned about data accuracy, correctness, completeness, and timeliness. Collecting data to this new high standard requires different job skills. Interpreting data using machine learning and AI (as part of the drive to greater autonomous operations) requires skills that are not resident in the company. Enterprise-level business intelligence for operations and for the head office did not exist in a business formerly operating as a collection of independent assets.

Boosting data acumen is now a foundational requirement for any company operating at scale in digital. As NRM has discovered, companies will need to invest in raising these data skills among employees and in recruiting the talent needed to leverage new data riches.

Downstream Companies

Every car owner has had direct and firsthand experience with the downstream petroleum industry—at the gasoline pumps. Globally, there are hundreds of thousands of fuel retailing businesses in a vast range of types: unmanned airports in the remote Australian outback, tiny parking spot refueling sites in dense cities, enormous chains like Shell, which operates 46,000 retail sites globally.[7]

In addition to retailing gasoline to the motoring public, downstream companies distribute heating oil, propane, diesel fuel, bunker fuel, and other energy products to manufacturers, farms, homes, truck stops, marine terminals, airports, rail yards, and construction sites. Supplying petroleum products daily involves fleets of trucks, rail cars, pipelines, and barges, and vast networks of tanks and storage assets strategically located close to markets and transportation hubs. Feeding the supply are small teams of energy traders that secure crude oil for the 690-plus refineries that produce these valuable products.

Case Study 6: VARO Energy

VARO Energy BV is a joint venture company owned by commodity trader Vitol Group and private equity firm Carlyle Group LP. The company operates a small refinery in Cressier, Switzerland, and has a 45 percent share of the Bayernoil Refinery in Germany. Products are distributed through a network of some fifty petroleum tank and blending facilities, and a retail network of a hundred-plus stations servicing markets in France, Germany, Switzerland, and Benelux (Belgium, the Netherlands, and Luxembourg).

FUEL RETAILING changes slowly because retailing works well at scale. My neighborhood Shell station closed during the pandemic for an overhaul, and I fully expected the new station to showcase the very latest in Shell's future of retailing—the part of Canada where I live has a very high penetration of EVs. Instead, the new station is indistinguishable from all the other Shell stations locally, except for that fresh-paint smell.

VARO Energy is illustrative of the many midsized fuel companies that supply mature markets around the world. The emphasis is on customer service and intimacy (sell side) and distribution logistics (supply side). The customer-facing side of distribution businesses is measured on margin, and the distribution side is measured on cost to serve. VARO has an additional element, trading, that provides the ability to optimize supply to the customer at a margin. Its moves are highly focused and deliver immediate impact.

BALANCE PRAGMATISM WITH INNOVATION

Small and midsized businesses must strike the right balance of meeting their annual economic goals with the need to stay market relevant. The fuels sector is intensely pragmatic—investments in the business must pay off, as there is little excess capacity in these businesses to absorb poor returns. In Europe in particular, the demand for petroleum products has been flat for a decade. Prices cannot be raised, volumes cannot be artificially boosted, and sales cannot simply aim at higher margin products, because these markets are so competitive. As IT Manager for Digital and Business Solutions Daniel Cadete, from VARO Energy, observes: "We want to invest in digital where it makes money."

As a result, VARO's digital program is twofold: capture any available margin opportunities that are accessible in the market, and help improve the existing business at all levels to become as efficient as possible. This is challenging in mature businesses that have been in operation for decades—most inefficiencies have long since been extracted from operations, and business practices have been in place for many years.

This emphasis on pragmatic outcomes translates into more demanding business case tolerances for changes, and lower risk appetite, which also yields quicker results that are customer relevant and market competitive. Conversely, such a model can produce copycat innovation and a dwindling ability to achieve breakout creativity over time. Innovative longer stride and less proven technologies like blockchain solutions are viewed as simply too risky. To compensate, VARO also maintains a small central team whose role is to explore these innovations with much less strict targeted outcomes and a much higher tolerance of risk.

In oil and gas, pragmatism trumps all else, and balancing mechanisms to counteract are required.

RUN THE TRANSFORMATION PROGRAM FROM THE TOP DOWN

With a highly pragmatic culture, companies find that a digital transformation program works best when it's driven from the top down. Bottom-up digital efforts can find themselves misaligned with what bosses have signed up for and lacking in organizational support and resources.

VARO has found that managers with clear line accountability, measurable expectations, and control of resources are at an optimum point from which to drive digital innovation. At this level of the organization, managers craft annual plans, set out their business improvement agendas, allocate their resources, and commit to delivery of results.

In recognition of delivery of that accountability, they achieve greater bonuses and higher compensation. Bottom-up ideas may be valuable, but unless the support for the business case is shared with a manager, the change is unlikely to be as successful as one that is obviously linked to plans and compensation. As Daniel puts it, "Start small and grow big. Do less to a higher degree. Do not overpromise. Take small steps."

One example is VARO's planned adoption of a customer-facing sales portal. Online ordering of many products is commonplace thanks to Amazon's transformation of retailing, but it is still very novel for fuel products. Such an idea (online sales of fuels) could have originated with the frontline, but the adoption is so disruptive that it is unlikely to gain much support without strong leadership. Online sales and direct customer access to place orders eliminates phone calls and email exchanges, transforms sales desk performance metrics, risks customer churn and dissatisfaction, and converts employees from their customer-facing roles to website managers.

Good ideas are valuable regardless of where they originate, but successful adoption works best top-down.

CREATE AN INNOVATION TEAM

Harnessing innovation even in a small business takes time, energy, and resources. For small and midsized businesses, too much innovation, in the form of too many small pilots and trials, one-off product experiments, and single-serving software products, can drive up costs and create a support nightmare. Too little innovation, or innovations that are too small, and the company risks stagnating or becoming uncompetitive over time.

VARO first trialed innovation in a distributed and decentralized structure. Any idea was considered, but the results were not sufficiently impactful, and the range of changes became unwieldy. To achieve greater impact from its digital investments, the company created an innovation team drawn principally from its IT group. This team takes on projects that entail higher-risk or less-certain commercial outcomes, and has responsibility for 20 percent of the total innovation budget.

To keep the innovation group focused on results, the company also created a digital steering group jointly led by the CIO and the COO. The steering group meets every two weeks to drive accountability and has line of sight to all digitally enabled change. The executive sponsor for digital innovation is the COO, who reports the results to the board.

The rhythm of a biweekly session is important to drive accountability. The steering group methodically reviews each project to check its status, entertains new proposals for change and allocates budget, balances the portfolio of projects across the business, frees up needed resources, and addresses any roadblocks or organizational misalignments that need attention.

Ask for innovation and you'll get it—but it might not always be what you wanted.

PREPARE FOR AN ANALYTICS FUTURE

The future of downstream appears slanted towards strong analytics.

For the past two decades, through the wave of ERP-enabled change, business cases were oriented around tangibly reducing easy-to-find costs—put another way, headcount reduction. Technology support (IT) reported to the CFO so that technology costs did not get out of hand, and no business case proceeded without a cost component. At the same time, costs can only go down so far, and in competitive industries, you can't cost save yourself to prosperity.

The commercial side of petroleum is very different because it is focused on margin capture. Upside margins are potentially unlimited. We can see this clearly in the occasional price spikes that appear in energy markets around the world. If you have the supply when the price is high, or you can figure out creative ways to satisfy demand for fuel that is, for whatever reason, price inelastic, then margins can be stellar.

To illustrate, in the aftermath of Hurricanes Katrina and Rita in August and September of 2005, oil refineries on the Gulf of Mexico shut down, removing 8 percent of US refining capacity for months and creating a temporary shortage of diesel fuel, which was badly needed for clean-up and rebuilding.[8] Margins were so strong that North

American refineries along the east coast that could supply the market made as much money in a year as they had in the preceding decade.

The challenge, however, is that margin-based business cases lack the hard tangibility common to cost-based business cases. As a result, deeper analytics are required.

Anticipating this future, VARO has invested most heavily in improving data quality and nurturing citizen analytics. It has become apparent that combining business domain or commercial expertise with analytics know-how is one of the keys to quickly sorting through available data and determining margin capture potential. The alternative, a stand-alone analytics team of data scientists, lacks the speed necessary to act on fast-moving commercial markets.

Related to the rise of analytics is another phenomenon: the mass customization of job roles. As the individual employee extends their analytics capabilities, they tune their job to the analytics that they have crafted. Analytics may become the Excel of the next decade— widely available but desk specific, and, with all the concerns about Excel and its lack of documentation, rife with hidden knowledge and poor maintainability.

Data analytics—it's going to be big.

Integrated Companies

Integrated oil and gas businesses are complex animals, with features that are common to many industries. They are stewards of capital, executing projects in common with EPC firms. Like dedicated trading houses, they buy and sell commodity products. They retail products to end consumers, in competition with other retailers. They run manufacturing plants that convert raw materials into valuable products. They are logistics businesses obsessed with assuring uninterrupted energy and chemical supplies to industry. And they are commodity businesses focused on resource management.

In short, they are perhaps best approached not as monolithic businesses but as collections of specific business models united in providing secure energy supplies sustainably.

Case Study 7: Repsol

Repsol is a Spanish energy company with global operations that conduct exploration, production, refining, and product sales in nearly
a hundred countries worldwide. Launched as a brand in 1948, Repsol has benefited significantly from its roots as a Spanish business,
enabling somewhat easier access to Latin American, African, and
Asian markets that operate in the Spanish language. Access to capital
markets enabled growth by acquisition, through such brands as Tal-
isman Energy (a Canadian transnational).

In addition to a compelling track record of oil discovery, the
company is distinguished because of its first-mover commitment
to achieving the demanding decarbonization targets enshrined in
the European Green Deal. It sees digital innovations as integral to
achieving this demanding goal across its four broad lines of business—
exploration and production, industrial (refining, petrochemicals,
trading, transportation, LNG), commercial and renewables (low carbon power generation), and consumer (retail stations).

Shortly into the twenty-first century, Repsol recognized that to
control its destiny and contribute fully to Spanish society, the company needed to invest more in R&D. From this arose what is now the
Repsol Technology Lab, whose role is to conduct the kinds of lab
research and analysis important to energy companies, including those
in the digital area. Since 2017, the entire company has been pursuing
an agenda to bring digital innovations to bear on its business, and
has amassed considerable experience in digital change. To do this, it
followed four key practices.

I WAS invited to join a panel discussion hosted by Repsol at their
Technology Lab facility outside Madrid in January 2020. The topic
was the application of digital technologies in the industry, and
specifically the role of blockchain solutions. Repsol had recently
invested in Finboot, a blockchain middleware startup, and blockchain was viewed as part of the solution set to helping Repsol
meet its goals in line with the 2015 Paris Agreement. Before the

panel got started, I leaned over to the Repsol panelist beside me to congratulate them on their environmental stance. She replied, "We have no idea how we're going to achieve this goal—but we are going to achieve it," echoing John F. Kennedy's commitment to America's Apollo lunar program in the 1960s.

REQUIRE DIGITAL INVESTMENTS TO BE ECONOMIC

Like many industrial companies, Repsol expects its digital investments to generate an economic return, but, more importantly, the returns are baked into the financial targets for the company. Once included in the financial plan, these investments are then far less likely to be cut from the budgets should commodity markets take an unfavorable turn, or should a pandemic strike.

The standard playbook for commodity companies is to trim discretionary spending in the face of declining prices for commodities. Typically, investments that are tied to hard capital (such as new plants and equipment), which are committed through purchasing or delivered on long contracts, are left more or less intact. Repsol views its digital plays not as discretionary but as must-do projects. In fact, some of its digital investments, such as those that convert capital ownership into subscription- or usage-based commercial structures, are likely to be even more economic in a downturn or pandemic situation. "We decided that all our innovation programs need to have a significant number of projects with a clear return on investment. If not, they suffer a lot," observes Tomas Malango.

Repsol makes its digital investment decisions by business area, with capital expenditure targets and cost and revenue benefits. The four business units (upstream, industrial, renewables, and customer) set different targets because their contexts vary. During the pandemic and related commodity market upheaval, Repsol maintained the bulk (80 percent) of its digital investments, including those aimed at bolstering its digital foundations (cloud computing, platforms, blockchain).

TAKE A STRATEGIC VIEW OF DATA

As part of its digital agenda, Repsol now takes a strategic view of its data, since digital is about enabling the capacity to grow, digital is all about data, and data is the main asset that digital unlocks. Almost 70 percent of its digital initiatives involve data, analytics, and AI. Data is framed as a pillar alongside people and process:

- We people are living in a digital world, and we need to improve our skills as digital citizens to be better professionals.

- Process efficiency is accelerated by digital innovation, which reduces energy consumption, and, in turn, our carbon and methane emissions.

- Data creates new intelligence from which we can capture value in a changing world.

Like many industrial businesses, the company generates tremendous quantities of data from its processes (oil and gas extraction, refining, product retailing). It is from these data holdings that the company can extract and monetize value. It is not always immediately obvious what new intelligence can be taken from the data, and the value may well vary across business units. In the case of the upstream and manufacturing units, safety and efficiency (both process and human) have been the dominant drivers of data analysis. In retailing, the focus has been on understanding the customer and their consumption patterns and buying behaviors.

This data orientation has important links to a broader message for Repsol's talent. Being seen as a digital leader in oil and gas is viewed as positive for attracting the next generation of talent, particularly in the new data-centric roles of data science. Competency in working with data is now a common feature for most jobs.

To accelerate its data acumen, Repsol kicked off the Data School project, in collaboration with a leading digital business school, to provide its employees with training in data areas. Employees in jobs that require deeper analytic capability take re-skilling courses to upgrade their performance. In addition, the school provides training in the kinds of new jobs that data, analytics, and AI require.

TRIAL NEW BUSINESS MODELS

Nothing is more likely to keep a board awake at night than the prospect that some new business model has been percolating away, unremarkable and imperceptible, until it's ready for market and experiences breakaway growth. For example, Uber entered eighty-four markets in just twenty-four months. Transportation services companies such as taxis were devalued overnight, wiping billions from balance sheets everywhere.

Repsol, in its position as the dominant market player in its domestic market, is in a position to identify, study, trial, and then roll out new business models. It has flagged, for example, that the relationship between energy provider and end customer is now in a position to change dramatically. End customers have access to more data about energy markets and offers than in the past, and they expect their energy services companies to be fully digital and to offer more embracing and inclusive services. In fact, the energy customer of the future might not even be a person but a machine or household.

The commercial model between energy supplier and customer is also up for change given digital's potential. Instead of a customer buying a tank of petroleum for their vehicle, could a customer purchase transportation as a service, available on demand? Instead of visiting a fuel retail site, could a vehicle owner purchase electricity of a specific quality (such as green energy) from any charge point available?

New business models are not limited to rearranging the assets and their owners in the energy value chain. It has taken a few years of trial and error, but big energy firms like Repsol are unlocking new commercial models for engaging with large technology companies, small startups, universities, incubators, and many other market participants.

SET UP CORPORATE VENTURING

Repsol is one of a handful of companies in oil and gas that have set up corporate venturing funds whose purposes are to invest strategically in promising new technologies (there are many such funds, from Saudi Aramco, BP, Shell, Chevron, ENI, ExxonMobil, and Equinor), such as clean tech, new materials, digital innovations, and energy solutions.

These funds are freed from the tyranny of winner-take-all financial goals, short-term ownership cycles, and immediate cash flow targets typical of more cutthroat traditional venture capital funds. Corporate venture funds usually have a single investor (the oil and gas company), which simplifies decision-making and provides a clear investment horizon. Such funds are often isolated from the day-to-day economics of the corporate backer and can continue to invest even when oil and gas prices have fallen.

Besides the possibility of an eventual payoff on exit, Repsol benefits in myriad ways from their strategic investments in digital, clean tech, and new energy businesses through the venture fund. The company gains exposure to fast-moving technologies that might struggle to gain traction inside a large corporate structure. They learn how to work in an agile way with startups. There are opportunities to apply the solutions in which they invest directly in the business. They gain experience with new technologies, scaling up, localization, regulatory needs, and new business models. Their brand improves among the young talented entrepreneurs in society. "We have integrated a lot of startups for internal projects, which is something ten years ago you could never imagine," notes Tomas Malango.

Startups apply for admission to Repsol's accelerator program, and a select handful obtain funding support. They are then modestly relieved from their own tyranny of constantly chasing the next round of funding. The startups can tune their solutions to improve their success (only 1 percent of startups are successful, but that rises to 30 percent when the startup is actively mentored by a corporate customer).

Repsol's digital funding is in the hundreds of millions of euros per year, the majority of which is focused on the low-carbon economy, digitalization of the energy industry, and the circular economy. A small proportion is aimed at enabling technologies such as cloud computing and blockchain.

EPC Services

With annual capital expenditures comfortably in the multiple hundreds of billions, the oil and gas industry relies extensively on the global consulting, engineering, procurement, and construction (EPC) companies to deliver capital both cost effectively and on schedule. EPC firms distinguish themselves based on the various dimensions of the energy industry where they achieve scale and prominence, and on the range of services offered, from **turnkey** project delivery to build-own-operate models.

Case Study 8: Wood

Wood PLC, based in Aberdeen, Scotland, is one of the world's largest full-service engineering companies, covering concept studies and pre-**FEED** activities, through project execution and facilities operations. Based on its roots as a leading fishing company that moved into supplying the North Sea offshore industry, it now principally services the oil and gas industry. From those early beginnings, the company has grown through a handful of major acquisitions and is now positioning itself to help deliver future energy solutions.

Service companies like Wood are now well along in embracing digital, and have perceptively accelerated these moves during the pandemic. Wood's transition is a good illustration of the kinds of moves that large companies can make to change themselves for the future.

INVEST IN A TRANSFORMATION PROGRAM

Like many large industrial concerns with a huge geographic footprint, Wood needs to communicate a consistent internal message about the future. That message is, "In a changing world with changing expectations, the company must remain relevant to its customers in order to preserve its competitiveness." Digital innovation is not positioned as the strategy itself, but rather as an enabler that is woven throughout the strategy. Adopting digital tools internally helps reduce cost and improve operational and project delivery efficiency, which is key to cost competitiveness. Leveraging digital tools externally to deliver results to customers helps the company maintain relevance in an increasingly digital world. Collaborating with digital partners who are leaders in their respective technology fields helps Wood avoid redundant internal investments, particularly where partnering provides a more sustainable and valuable technology proposition to customers.

Messaging is very potent when it is backed up with investments and action. To propel the strategy forward, Wood has created the Future Fit program, consisting of a series of transformation streams of work that tackle different dimensions of change, balancing both internal and external shifts.

Future Fit will likely change over time as various streams reach their logical conclusions and other change needs come into view. For example, one Future Fit stream focuses on the skills required in a future workforce. At one time, digital coders were in high demand for their skills, along with cloud computing experts and blockchain **full stack** developers. However, creator and developer tools have themselves embraced digital innovations, becoming so much simpler to use that anyone can deploy them. These low-code, no-code platforms remove the excuse that digital is somehow too hard to learn or not suited to an industrial environment. Future Fit is helping guide recruiting so that preference is given towards new hires who are naturally curious and flexible continuous learners, willing to experiment, and collaborative problem solvers.

Funding a named initiative, and supporting it through a concerted multi-channel communications effort, is a powerful way to drive the change message forward.

CREATE AN INNOVATIVE SOLUTION LAB

Increasingly complex challenges need a cross-functional and multi-specialist approach that is often not available through a single vendor. To stay relevant and competitive, Wood has developed the concept of a co-lab, a solution development experience that brings together consulting engineers, customers, and technology providers in a creative and unconstrained place to collaborate. These labs offer

- an opportunity for clients and consultants to ideate around a complex problem in need of creative solutions;

- a forum for managers to solve internal performance challenges;

- a place to work differently, with agile methods and techniques;

- an experimental zone to explore new digital technologies and their application in industrial settings;

- a chance for employees to be innovative and curious about the art of the possible, and to learn new techniques and technologies;

- a promise of breakthrough gains in performance, quality, and cost;

- a setting for the invention of completely new digital solutions, products, and services; and

- a relationship-building experience to help form new, deeper, and more impactful relationships with their customers.

The co-lab is an ideal place to foster the application of agile methods, which lend themselves to the exploration of problems and iterative solution development, and lead to valuable real-world outcomes. Candidate problems where agile applies well include those featuring uncertainty of outcome and few hard and fast rules to follow. Classic engineering problems, such as **pressure vessel** design, with its well-defined mathematical models and precise solutions with known parameters, appear less suited to agile approaches. Co-lab sessions are fantastic nurturing grounds for innovative ideas that subsequently feed solution development using agile methods with the Digital Factory.

BUILD A DIGITAL FACTORY

How do you strike the right balance between allowing customer-facing teams to develop clever but highly specific solutions as one-offs versus creating a central technology team that is solely responsible for innovation delivery and becomes a bottleneck in the process? Wood's answer is the Digital Factory, a central team whose role is to support and scale innovation at the "edge," another name for a client-facing project. The Digital Factory ensures that a product life-cycle approach is taken to digital development and that enterprise security needs are addressed. Consider the case of an industrial client who sets forward a requirement to optimize battery recharging for a fleet of city buses. Democratic digital tools make it very easy for such a solution to be delivered directly by the client team as an edge solution. On the other hand, such a solution, if delivered as a product and not a bespoke one-off, can be architected and branded with a view to much wider application to bus services globally, and to many fleet operators.

Enter the Digital Factory, which helps develop and evolve the solution to consider the wider range of issues that many customers likely have, with an architecture for multiple customers, that can scale up and integrate with a wider range of third-party systems. The edge team then takes the factory solution back to the client for direct deployment.

Only a limited number of solution ideas that feature powerful scale effects or genuinely global appeal get taken up and funded by the Digital Factory as enterprise solutions, leaving the edge teams with considerable autonomy to serve clients directly.

The Digital Factory is structured to behave like a startup in the technology industry, using agile methods, product managers, and scrums, which allows Wood to sharpen its use of these tools and to train its people in their application. In a relatively short time, the Digital Factory has been instrumental in helping produce a roster of new product candidates in such areas as maintenance, spare parts management, noise analysis, project ESG compliance, and emissions monitoring. These products will deliver their own revenue streams, but may one day even be monetized as spin-offs.

Consider, for example, the challenge of forest fires. Arboreal fires can have a devastating impact on oil and gas infrastructure. Traditional methods for monitoring fire activity include expensive manned watchtowers and regular helicopter overflights, which, at best, cover only the visible terrain. The arrival of inexpensive and near continuous satellite coverage now generates a steady stream of imagery to analyses, but outstrips the human capacity to inspect the imagery for fire activity. A Wood edge team identified the opportunity to apply machine learning interpretation of images, and from this developed an app that interprets the images and provides continuous, 100 percent coverage of changes in the forest canopy. Expensive manned flights and the associated fuel emissions are a thing of the past.

RETHINK THE CUSTOMER RELATIONSHIP

An enduring challenge facing the EPC sector is in the structural relationships in the supply chain between itself, its clients, and its suppliers. These arrangements have withstood the ravages of the commodity cycle, the waves of technology change, and even the pandemic. As Azad Hessamodini notes, "The win-lose mentality has to change. The contracting methodology and standards, norms, KPIs, are all based on win-lose."

The current relationships cannot be described as win-win. Clients have little or no incentive to share in the upside of innovation and do not require innovation. Performance measures artificially create competition among those who should be collaborating. Some perverse measures can even encourage the wrong behavior, such as using outdated manual labor methods that boost the contractor's revenue rather than leveraging cost-reducing technologies that improve productivity for the contractor (but which reduce the contractor's costs and absolute revenues). Commercial arrangements negotiated through procurement departments are often simplistic and unfit for tackling complex challenges for gain- and risk-sharing. The pendulum of contracting methods swings from reimbursable contracting to fixed pricing and back again. Sophisticated buyers realize that breakthrough advances are achieved through collaborative teamwork, which necessitates an equitable allocation of risk and reward.

Data continues to be at the heart of the relationship challenge. The industry's culture of treating all data as proprietary, a lack of data sharing and transparency, few standards, and prohibitive contracting language actively inhibit digital adoption, slow down digital innovation, and contribute to confrontational relationships. While some data (reservoir and resource-related data, for example) could be commercially sensitive and central to an operator's competitive advantage, other data related to facilities operations are no longer proprietary and can benefit from democratization to enable machine learning and AI advances for the benefit of the industry as a whole.

However, digital tools and business models are starting to unlock the value trapped in the structural shortcomings. Using the co-lab as an ideation platform allows Wood to surface these challenges, and for the parties in the lab to explore alternatives to the current confrontational business model. The lab in turn provides actionable insights for the Digital Factory to help reframe the commercial model.

Not every client is ready for this discussion—it takes a certain level of maturity and sophistication to accept that the status quo is untenable and that more equitable possibilities exist.

Case Study 9: Worley

Worley is a global provider of professional project and asset services in the energy, chemicals, and resources sectors headquartered in Sydney, Australia. Worley has consistently grown through acquisitions of key services (such as environmental services and **fabrication**) and companies in targeted geographies (including Colt Engineering in Canada for exposure to the Canadian oil sector, and S.E.A. Engineering for the offshore). The company now has a robust footprint of asset-related services around the world and represents the challenge of aligning and coordinating a complex, matrix service–oriented business model of engineering disciplines, industry sector and projects, and geographic locations.

All companies, to one level or another, must react to the combined impacts of energy transition, digitalization, and ESG, and they frequently need assistance to do so. To prepare itself for the future, and to support the looming industrial demand for help, Worley undertook its own internal transformation. Under a new CEO, Worley changed its operating structure, focused on devising digital solutions, and launched a new purpose statement—to deliver a more sustainable world.

The COVID-19 pandemic has only served as an accelerant, advancing change at a pace internally estimated as being five times faster than expected. The workforce has swiftly embraced remote working and online collaboration, and digital and sustainability quickly became core to the company. Several aspects of its digital change agenda stand out as leading practices.

HAVE A POSITIVE, INSPIRING NARRATIVE

Worley believes that the scale of change required to build a low carbon future can be achieved only through digital enablement. The infrastructure required, in light of the available skilled workforce, congested cities, and the legacy asset mix, will need the efficiencies created through digital.

The changes are framed as positives—sustainability and energy transition will require one of the single largest reallocations of hard investment capital in history, a dramatic growth opportunity for those in the engineering world. Trillions in capital will be spent to produce, distribute, and consume new energy in the form of new technology, new assets, new business models, and new ways of working. Digital, as the key enabler for energy transition, is no less of an investment category in its own right, and becomes less of an internal project around which transformation effort is spent and more an expression of how work gets done.

This messaging—positive, growth, inspiring—is the same for internal stakeholders across all lines of business, geographies, and industries, as well as the external market of communities, customers, suppliers, capital markets, and regulators. Worley's new CEO had a ready platform to articulate the changes and drive the communications at an enterprise level.

MANAGE GLOBALLY, EXECUTE LOCALLY

As a complex distributed organization, Worley needs to balance the needs of the enterprise for coherence in its approach to digital, sustainability, and energy transition with the inherent creativity that is fostered by customer-facing teams. Too tight a control from the top risks killing off innovation by making financial hurdles overly stringent, slowing down decision-making and restricting access to capital. Too loose a control, and capital can be needlessly wasted in mundane undifferentiated areas such as cloud, cyber, and networks. As John

Pillay notes, "Digital works best at scale... You can't have a free-for-all on everything."

Both digital and energy transition are represented at the Group Executive. Its importance is a clear nod to the corporate strategy and brings focus to both the external market-facing opportunity and the internal digital transformation. It is also an explicit recognition that the changes are intended to be strategic and transformative.

To capitalize on the creative energies of its thousands of engineers, local teams, projects, and verticals are largely free—and very strongly encouraged—to experiment and exploit digital tools throughout the business. Regional champions take care of promoting the transition locally, while an innovation council helps nurture change and popularize promising developments. An innovation hub helps surface ideas from the field, an explicit acknowledgment that digital innovations have been widely democratized.

Digital teams collaborate with industry experts to design roadmaps to guide customers towards their decarbonization goals. Example outcomes include data to optimize the operations and maintenance schedules of offshore wind farms, robots to carry out hazardous work, and platforms to accelerate engineering design scopes to be faster and cheaper in bringing large projects online.

ACTING ON one employee's innovative idea, Worley has developed a robot that can identify and collect **catalyst** and other toxic substances from inside vessels during maintenance, repair, and turnarounds. Vessels are inherently dangerous work sites, as they are enclosed, can contain remnant vapors, and lack convenient egress and communications.

As an example of the balancing act, consider data. Specific engineering projects generate islands of rich engineering content and asset data at a local level, which could also prove to be valuable enterprise assets if managed differently. Local project teams gain little project benefit, and potentially considerable incremental cost, from treating such data as an enterprise asset.

SUPPORT YOUR PEOPLE

At Worley, the argument is well and truly won that change management is at the heart of a successful transformation. The company backs up this view with a robust annual investment in securing the hearts and minds of its people on the imperative for change and the future of work.

As an engineering and project delivery organization whose success is based on the talents of its people, Worley capitalizes on the natural tendency of its people to want to stay abreast of the latest developments in their respective engineering fields of discipline. The company pioneered the concept of a digital passport, a kind of digital literacy program that features many small quick courses, events, and learning conferences, the attendance and completion of which earns various badges of recognition, gamifying an education program. Just three months after launch, twelve thousand people had achieved some level of digital accreditation, almost a third of the global office-based workforce.

The company organizes frequent internal virtual and global conferences, bringing its people together to explore digital and energy topics, celebrate successes, and transfer knowledge and practices across its geographies. Up to a thousand join in these fast-moving, focused virtual events, which are organized along regional lines in respect of time zones, punctuated with the occasional global event. Unsurprisingly, it is considerably more persuasive to adopt a novel idea or solution that has been proved in another regional team, project, or office than to embrace an edict from the corporate center.

MEASURE IT TO MANAGE IT

Worley requires that every new digital investment have a clear line of sight to benefits and tracking mechanisms, which is one of the differences between *doing* digital and *being* digital. Perhaps reflective of its engineering heritage, Worley has embraced various tactics to objectively measure its progress in adopting digital innovations. At any one time, the company has tens of thousands of individual projects underway, each of which is measuring the impacts that digital innovations are having and accumulating a valuable dataset for identifying early winning concepts and applications.

One of the first measures Worley employs is the forecasted impact that a digital innovation is expected to have on any of the key corporate targets, particularly people utilization. Freeing up staff to carry out higher-value-added work is economically impactful. The nature of the impacts is captured—specific time savings, value of quality improvements, and reduction in cost.

Lack of funds is not an excuse. Managers can deliver online training, conferences, meetings, project reviews, inspections, and a host of other project- and office-related activities virtually and for very little, if any, incremental cost.

Harder to measure items, such as the potential value of new services, or the impacts of new business models, will be addressed in time. Eventually, digital innovations become expected, as do ways in which work is done, and are so thoroughly part of the culture that an individual project emphasis on digital is no longer viewed as exceptional.

Finally, Worley is trialing measures that seek to bring digital, sustainability, and energy transition together. As an example, the company has created a carbon emissions index for projects and proposals to help quantify the potential environmental impact.

Key Practices

The case companies offered many tactics that have helped them achieve significant success. I have grouped these by broad category, such as crafting the best narrative or executing with purpose. Within this discussion you may find some fresh ideas that help you with your digital journey.

Frame the Narrative

As leaders you will need to articulate a compelling narrative or story that explains, inspires, and motivates your stakeholders to action. How do you frame the narrative to explain your digital agenda?

MAKE IT PURPOSEFUL

Create a single narrative for all stakeholders and tune it slightly for customers. Make it simple so that it's authentic. Focus the narrative on the problem you are trying to solve and on specific targets, such as faster and better decisions, improved reliability, more efficiency, less waste, and increased profitability.

Explain how the use of modern tools can create a platform for growth to fuel acquisitions, enable a more flexible lifestyle, and prevent a customer from considering backwards integration. Frame the digital change as a growth opportunity, not a survival strategy.

MAKE IT REAL

Focus on outside forces such as the need to navigate the two mega trends—energy transition and digital transformation. Position the change as the need to protect your hard-won market position. Introduce the idea of transforming legacy assets so that they can perform like premium assets. Focus on process alignment and tools standardization that have real impact on daily work.

Share the brutal truths about how your company is performing in the competitive capital markets. Take advantage of a change in leadership, or acquisition, to launch a digital drive in connection with these brutal truths.

MAKE IT ABOUT THE FUTURE

Focus the narrative on the future, while acknowledging the past (your employees are vested in the successes of the past). Weave your digital strategy into your company strategy (the world is changing, and we must change; everything is "becoming digital," so must we). Articulate a vision of the future that speaks to the future workforce, work processes, and commercial models. Describe the "operator of the future" that maximizes performance without being present on-site. Articulate a new operating model that incorporates remote operations, autonomy, and digitally enabled operations.

MAKE IT ABOUT THEM

Position digital innovation as helping relieve employees of tasks better done by machines, and letting people focus on human-only work. Combat the fear of losing a job to technology by explaining the need for people to do more troubleshooting, not work planning.

Challenge employees by asking them to imagine being owned by a company in a different industry, and ask what the new owner would want to change.

MAKE IT RELEVANT

Link the narrative to the culture norm set that the company is creating. Link the digital strategy to the corporate strategy (or business unit strategy or functional strategy).

Create a vision of "The New Well" (or rig or asset) that incorporates the possible innovations to dramatically accelerate delivery, lower operating cost, bolster utilization, reduce the environmental footprint, lower abandonment obligations, and minimize waste byproducts.

Create a Governance Structure

A digital change program requires structures and mechanisms to drive its execution. How do you govern digital execution?

MAKE LEADERSHIP ACCOUNTABLE

Create a digital steering committee, composed of CEO, president, digital leader (or other relevant structure, such as business unit leader, head of sales, and head of digital). Create a digital leadership team (for example, composed of the CIO and COO) to achieve the right balance of technology and domain expertise.

Assign digital accountability to the executive committee. Promote board-level dialogue about digital change and impact on the company and the industry.

Hold managers accountable for driving a digital innovation into their business (embed it in the plan, ask them regularly how they're doing against their plan, reward their successes, and sanction their shortcomings).

BUILD CROSS-COMPANY LINKAGES

Link the chief technology officer to a VP responsible for digital in the business unit (a dotted line relationship). Create an operational technology committee, with two data-savvy non-executives from across the business, including the field, to help promote alignment for implementation.

ORGANIZE FOR SUCCESS

Move the responsibility for digital, IT, and OT from the CFO to a different executive leader. Merge IT, OT, and digital together under one leader to resolve conflicts. Place overall digital leadership under the operations leader (director level). Create a digital center of expertise or excellence. Create an innovation center with an experiments budget, lower business targets, and higher risk tolerance.

Set up a combined co-located team of data scientists, IT (software, architects, and data lake), and business professionals (subject matter experts) to work together on digital initiatives.

LOCK IN SPONSORSHIP

Secure consistent and unwavering support from the CEO for the digital change, and seek continuous reinforcement of digital changes. Create time for the digital executive sponsor to attend digital experiments to keep up the pressure, unblock issues, and maintain energy. Demonstrate relentless overall executive support for the digital changes to reinforce the expectation that digital adoption is the path forward. Give digital project experimentalists full executive support so that a failed project will not feature on year-end reviews or impact bonuses.

ADOPT LEAN MANAGEMENT PROCESSES

Create a guiding program for the change and provide it with differentiating branding. Divide the program into themes and streams, including internally focused changes as well as externally facing changes (aimed at customers, suppliers, and other stakeholders).

Set aside a budget for digital experiments, divided into capital and operating, under an executive sponsor. Divide digital efforts by maturity—edge (or experimental prototype), emergent (solid proof of concept), and enterprise (scaling up). Aim to get a digital change into production and set aside scaling up so that scale issues do not derail trials.

Launch a strategy desk to coordinate strategic moves, including digital innovations and experiments. Define boundaries for innovation, with tighter limits on enterprise architecture and tools and looser controls on project-specific tools to trial.

Engage with the field to reduce the resistance to change. Give the field a role in capturing data and empower them to act on insights from the data.

CREATE SIGNATURE DIGITAL EVENTS

Hold weekly scrum meetings. Include business professionals with a financial focus on any digital or technology transformation teams to maintain attention on cost, ROI, etc. Include business professionals from sales and marketing to maintain focus on customer and market acceptance.

Introduce one agile technique for the executive team to adopt, such as a scrum-style meeting.

Celebrate any digital change victories that reinforce commitment to the journey.

Communicate Progress and Results

A wide range of stakeholders will be interested in what you're accomplishing. How do you communicate your progress and achievements?

GET THE RIGHT CONTENT

Create a presentation on the goal and vision for your overall digital program, and organize a specific campaign to deliver the presentation to as many internal audiences as possible. Over time, de-emphasize digital as something special and instead encourage digital in everything you do. Communicate why digital works and not just the results achieved.

FIND THE SPOKESPERSON

Target the key influencers in teams who are able to move others to trial and adopt innovations. Have the CEO or other executive sponsor occasionally speak about digital successes.

FLEX THE FORMAT

Send out updates about digital progress, tuned to different audiences, on a frequent and regular basis. Report results of digital experiments quarterly to executive team and full staff. Hold live executive talking sessions. Hold cross-company open forums to discuss solution proposals and ideas, showcase innovation ideas, and demonstrate technology. Hold an annual open mic session (small groups of eight) to encourage intimate discussions.

Run a frequent series of internal webinars or conferences to showcase technologies, solutions, pioneers, successes, and misses. Organize a regular digital town hall briefing for all employees, and focus on a digital outcome or solution.

LEVERAGE TECHNOLOGY

Use a multi-channel approach to communicate the change agenda, using available company services such as Yammer and Slack. Use the company intranet to post stories about digital innovations. Use LinkedIn and other relevant popular social platforms to post stories and articles about digital. Create a separate digital section on the company internal newsletter.

Create regionally organized virtual forums for employees to learn from each other, organized to maximize attendance. Create discipline-driven (such as engineering, finance, and supply chain) virtual forums for employees to learn from each other.

Use mechanisms to reach all workers (for example, some operators may not be permitted to have devices on person on the job).

Target the Opportunities

There is more opportunity to apply digital innovations to legacy businesses than there is budget. How will you target your opportunities?

SET UP FOR SUCCESS

As a precursor to analyzing any area or opportunity for digital innovation, first put the process through a continuous improvement exercise. Tie digital investments to some measurable impact, such as lowering costs, increasing revenues, and achieving compliance. Focus digital innovation on the problems that you have, not necessarily on achieving a specific target (such as growth or cost).

Create a simple one-page budget request, with a thirty-word description and a dollar cap. Appoint the role of an opportunity manager by department, product line, or line of business to surface the right opportunities to work on and the highest-value business problems to solve. Select a few large enterprise-level digital changes, and supplement with a portfolio of smaller changes that release value quickly and steadily to signal that digital is a journey.

Launch an innovation hub to solicit digital ideas from employees. Stress-test digital projects at double the cost and half the benefits, and cancel any that are negative. Identify the true bottom-line impacts of all digital efforts (no matter how small), and factor the results into monthly results and ongoing staff bonus calculations. Build a detailed and documented expenditure roadmap (three-to-four-year), and review with the board.

MAINTAIN MOMENTUM

Aim to accelerate on every dimension possible—time to decision, time to develop, time to test, time to deliver. Look for like-minded business partners willing to trial collaborations in the supply chain. Abandon any uncommitted partners. Leverage events (such as the pandemic) to drive change.

Encourage your employees to listen for and surface insights throughout the customer base and ecosystem to detect opportunities. Engage directly with customers, key suppliers, and technology companies to discuss digital innovation opportunities. Implement a revenue-sharing model with customers, based on commodity prices.

Develop Digital Capability

Since digital adoption is not a destination but more of an ongoing journey, you need new skills. How do you develop new digital capabilities?

SET OUT GUIDING PRINCIPLES

Be location agnostic in terms of where software talent comes from and where it lives. Learn agile ways of working from your IT team, who call it **DevOps.**

Develop skills in business process management (modeling, improving, streamlining). Map out a process before applying any digital innovations to it. Leverage vendors to help boost exposure to how other companies execute similar processes.

Identify leadership champions (in middle management) to drive change, and identify supervisor champions (on the frontline) to encourage others to change. "Explain and Train"—explain to employees why a change is valuable, and train extensively on the new tools and technologies.

ADOPT LEADING PRACTICES

Conduct an assessment to see where your organization stands on data maturity. Trace a data element of organizational importance (production volume) to see how the understanding of data changes through the organization.

Launch a digital literacy program, and gamify by incorporating a digital passport with badges and accreditations. Develop digital literacy at all levels to help bring people along for the journey. Invest in training in key foundational tools, such as SharePoint and Teams— videos, lunch and learns.

Leverage virtual video call capability to create a virtual training capability. Run multiple weekly small, brief (thirty-minute) learning events to teach the basics of digital. Provide development training on the soft skills side of digital (tolerance for ambiguity, change management, leadership) to create a more resilient workforce. Organize internal virtual conferences and webinars to showcase digital innovation and maximize participation.

Apply agile methods of working on digital innovations. Aim agile methods at non-precise, fast innovation.

Assemble Lasting New Skills

Over time, your digital change agenda will surface new skills that you need. How do you assemble those skills?

ADJUST SELECTION CRITERIA

Search for curiosity and flexibility in new hires, in addition to digital skills. Find people who think in a win-win fashion instead of win-lose. If you are growing organically, target any acquisitions based on the digital skills needed (sensor know-how). Change the hiring criteria for frontline supervisors to emphasize leadership attributes, since digital improvements are eliminating administrative work.

Attract data integration skills to take on the task of extracting data from various sources for use in analytics pilots.

SOURCE TALENT DIFFERENTLY

Outsource or get help in new areas (cloud, cyber, blockchain) rather than trying to develop expertise internally. Leverage external experts via subscription to monitor for cybersecurity.

SUPPORT TALENT DIFFERENTLY

Form communities of practice for new digital tools such as Teams, SharePoint, and SpotFire. Form communities of technical capability in such areas as IoT, cloud, cyber, and blockchain.

Create digital awareness training specific for frontline supervisors, as their role has likely experienced significant process change because of digital. Create on-the-job learning opportunities by changing work expectations to incorporate digital innovation.

Work with a supplier who knows how to do agile and copy their practices on an internal project. Bring in outside vendors to train workers. Teach usage with personal data (such as credit card data). Create citizen digital analysts who are fluent in new digital tool sets.

BUILD UP NEW SKILLS

Learn about how to sell data—pricing, marketing, pitch, and value. Work with frontline customer-facing staff to identify gaps in the market for possible digital solutions and pricing models.

Reduce the emphasis on applications as the center of budgeting and replace it with a focus on data. Build up data analytics skills internally, as they are in short supply globally and there is unlimited demand for analysis. Anticipate a bottleneck in data delivery skills (analytics, business intelligence, machine learning).

Identify niche skills (individuals who have intimate knowledge of internal systems, PLCs, DCSs) that bridge mechanical and ERP data sources in plants with new data lakes.

Equip your team with new no-code and low-code tools.

Work the New Ecosystem

A lively ecosystem of digital innovators and their supporters exists in most urban centers. How do you leverage your ecosystem of suppliers, customers, and other stakeholders?

BROADEN THE COMMUNITY

Broaden the ecosystem to include more technology companies and partners, universities, R&D houses, incubators, and accelerators. Broaden the ecosystem to include companies focused on ESG, environment, and emissions.

Collaborate with universities to run educational programs on abstract topics such as innovation. Leverage online and virtual training sources from the community instead of on-site training.

CHANGE YOUR RELATIONSHIP WITH SUPPLIERS

Fix procurement rules that prohibit or limit collaboration with suppliers on innovation. Learn to work with small technology companies. Actively work with small companies (co-authoring white papers) to help them get off the ground.

CHANGE YOUR RELATIONSHIP WITH CUSTOMERS

Engage with your customers to create a win-win outcome, more equitable distribution of value, and more sharing of risk. Create incentives to encourage customers to act in your desired model.

Create a collaboration lab (co-lab) for working with clients to carry out ideation, execute design thinking, foster curiosity, and deliver innovation. Create an innovation-sharing consortium with related but not competing companies, such as customers and international peers.

Execute with Purpose

Starting out with modest changes can build up the capability to execute larger, longer-stride changes. How do you execute with purpose?

CAPTURE QUICK WINS

Get rid of your manual internal processes to signal that you're serious about transforming your business. Target for remediation any manual processes that are brought in through any acquisitions. Craft your contracts to see that you get access to the data generated by your tools. Start a digital project only once you know how it is going to be measured.

GET TO THE CLOUD

Get everything into the cloud—data has to be accessible to be ready for a future of machine learning and AI. Move 100 percent to the cloud, including SCADA. Virtualize datasets from various applications (production data, financial data) to help become more predictive.

HARMONIZE DATA ENGINES

Move to a single SCADA standard. Focus on improving data quality so that once data is accessible it can be reliably used for automated decision-making. Move data from various sources (historians, fluid life oil analysis) to accelerate analytics and pilot projects.

LINK ESG AND DIGITAL

Enable ESG reporting exclusively with digital to drive interest and usage. Leverage digital solutions to understand impacts of carbon taxes on customers.

MAKE IT FUN

Hold an app launch party—a group download of a key app followed by teaching employees how to use it in a fun session. Re-architect work to take advantage of work from anywhere.

MAKE IT ECONOMICAL

Link digital outcomes to bonus schemes to create focus on their impacts and motivate adoption. Spin off internal innovations that could be readily productized.

KEY TAKEAWAYS

Digital is a contact sport. Leaders craft holistic change programs to drive their agendas. Here are just a few of the lessons from this survey of the industry:

1 Companies throughout the oil and gas value chain in all geographies are aggressively repositioning their businesses for a more digital future. There is no place to hide.

2 Digital innovations are not just tools for home office use. Digital is impacting operations as well as customers and suppliers.

3 Energy transition and digital transformation are unlocking an unprecedented opportunity for incumbents in the industry.

4 The pandemic has proved to be an unexpected accelerant in promoting the adoption of digital innovations.

5 There is no silver-bullet solution that quickly enables change to happen. It takes painstaking, unrelenting, and time-consuming effort to drive forward.

6 Engaging the whole of the organization, not just the early adopters and digital champions, is instrumental to success.

7 Change is very much top-led, with steering committees and leadership teams reporting directly to the CEO.

Notes

1   "McCoy Global Inc. Announces Strategic Asset Acquisition of 3PS Inc.," McCoy Global, January 4, 2017, mccoyglobal.com/news/mccoy-global-inc-announces-strategic-asset-acquisition-of-3ps-inc.

2   Bruce McCain, "The Facts Behind Oil's Price Collapse," *Forbes*, February 9, 2015, forbes.com/sites/brucemccain/2015/02/09/the-facts-behind-oils-price-collapse.

3   "Encana Reaches Agreement to Sell Bighorn Assets to Jupiter Resources for Approximately US$1.8 Billion," Intrado Newswire, June 27, 2014, globenewswire.com/news-release/2014/06/27/1023740/0/en/Encana-Reaches-Agreement-to-Sell-Bighorn-Assets-to-Jupiter-Resources-for-Approximately-US-1-8-Billion.html; Tourmaline Oil, "Tourmaline Completes Strategic Acquisition of Jupiter," Cision, December 18, 2020, newswire.ca/news-releases/tourmaline-completes-strategic-acquisition-of-jupiter-892628606.html.

4   "Natural Gas Prices—Historical Chart," Macrotrends, accessed June 29, 2021, macrotrends.net/2478/natural-gas-prices-historical-chart.

5   "Ovintiv Inc 2019 Annual Report," Ovintiv, February 21, 2020, filecache.investorroom.com/mr5ircnw_encana/871/download/2019-annual-report.pdf; "Ovintiv: Number of Employees 2006–2021," Macrotrends, 2021, macrotrends.net/stocks/charts/OVV/ovintiv/number-of-employees.

6   "Brookfield Infrastructure to Acquire Western Canadian Midstream Business," Brookfield, July 4, 2018, bip.brookfield.com/press-releases/bip/brookfield-infrastructure-acquire-western-canadian-midstream-business.

7   "Why Choose Shell Retail," Shell, 2021, shell.com/business-customers/shell-retail-licensing/about-shell-retail.html.

8   "Oil and Gas Disruption from Hurricanes Katrina and Rita," CRS *Report for Congress*, April 6, 2006, 14, everycrsreport.com/files/20060406_RL33124_c85587d5b537742f448395fdfea60a3fad64ae67.pdf.

CONCLUSION

ASEBALL WAS a childhood pastime of mine. My dad gave me and my brother a set of the oldest baseball gloves imaginable. In hindsight, they really belonged in a museum. He taught us how to throw and catch the ball and how to hit, and eventually I entered a short-lived career playing sandlot ball near home. I was a pretty good hitter, and I gravitated towards pitching and catching, because those two positions were the most active in the game. The unique twist in our version of the game was the frequent postponement due to fog, which was sometimes so thick that the fielders could not see home plate to know if a ball was coming their way.

The thing about baseball is that even when your team falls behind, there's always a chance it can catch up. The opposing pitcher flags as the game wears on, the hitters come to life, a rally gets going, and pretty soon some runs appear on the board.

I'm not so sure that's true of digital.

The world is littered with companies who thought they could catch up after some digital innovations began to appear: taxis, retailers, news outlets, entertainment. Only outfits at gargantuan scale, with resource depth and big turnover, look like they can play into middle innings and survive. Walmart is a case in point—they waited a very long time to match Amazon's online model, but they are surviving. Other retail stores? Not so much.

In May 2019, a conference organizer contacted me to discuss his findings from a recent trip to an oil town. He was considering launching a conference about digital transformation in the heart of oil country but was receiving a discouraging reaction. He floated his idea of the demand for such a conference to a number of large companies, and was politely told that oil and gas is "already digital." The game was already over.

At the time, it was accurate that some aspects of oil and gas were highly digital. But after a collapsed oil market in early 2020, followed by two years of pandemic mayhem, the worst global recession on record, and the dramatic shift to remote working by the industry, the sector has been forced to accept that the game was not over. It had barely started.

Fortunately, it isn't too late to get in the game. As governments vow to phase out hydrocarbon fuels by 2050, consumers adjust their demand preferences to other energies, and capital markets pressure the industry to change, there are still billions of engines that require fuel, millions of homes that need heat, and enormous industries that need chemicals. Digital can help make the next twenty or so years of oil and gas its most profitable, cleanest, and safest ever, leaving people, companies, nations, and the world in a better place to weather the transition.

To win in this digital game, oil and gas leadership teams must want to win and not simply strive for second place. Other industries have learned that in digital, winners frequently take all, leaving no room for second-place finishers.

Here are five signs that would convince me that the oil and gas industry has fully digitized and we're in the final innings.

Everything is digitized: Everything that can be digitized will be digitized.[1] In oil and gas, that includes the products of the industry, its tools and equipment, rental gear, services, data, vehicles, workers, and even rocks. The costs of using digital technology are rapidly falling to such ridiculous lows that cost is no longer a barrier.

Work is rethought: Oil and gas companies will design work for autonomy, and then add humans. Today, oil and gas continues to design work around a human worker, and supplements that human with technological tools to do the job (a laptop, a tablet, a mobile phone). Solid industrial digitalization means designing jobs for a robot or an AI engine, and figuring out the role of the human to improve the effectiveness of the bot. We can see how this model has completely transformed manufacturing. A survey of job postings in oil and gas suggests jobs are still very much human-first.

Digital achieves critical mass: The penetration rate of leading digital technologies in oil and gas will be 50 percent or greater. This means implemented by over half of oil and gas companies. "Implemented" does not mean "in pilot," or "being tested"—it means fully deployed in the business. The key digital technologies that must be widely deployed for the industry to claim that digital is done include cloud computing, bot technology, blockchain, virtual reality, and the IoT.

Digital oil companies exist: Oil and gas companies will brand themselves as "digital." Hard asset industries have consistently viewed themselves as immune to the impacts of digital innovation, but recent experience shows that this is a false premise. Brick and mortar retailing, banking, mobility, transportation, hoteling, printing—these industries were at one time based on converting shareholder capital into hard assets that were put to work in a business model. Every single one now has a digital market leader who competes directly in the sector, but with a fundamentally different and better balance sheet.

Oil and gas is rebranded: The industry will be fully rebranded back to its historical narrative as one of the most technologically advanced anywhere. For the past decade, since the Deepwater Horizon event in 2010, the industry has been on the defensive. The narrative is about safety, environmental respect, compliance, and sustainability, as the industry surrendered technology leadership to many other sectors (digital companies, banks, and telecoms). The sustainability narrative is important, but it is inherently defensive since the industry cannot reduce all of its environmental impacts to zero (combusting the product creates the final unsolvable environmental effect).

Oil and gas is a long way from being digital. Moreover, the industry is not going to disappear anytime soon. There is still untold opportunity for young people and tech entrepreneurs to find fortunes in the industry. The game is only just beginning.

Notes

1   Loucks et al., *Digital Vortex*.

ACKNOWLEDGMENTS

W E ARE forever grateful and indebted to a huge range of contributors to the development and production of this important book.

First, our deepest thanks to our friends and contacts in the global oil and gas industry who donated their time, and secured their organizations' support, to contribute their case studies to the book: Shawn Allan, Cory Bergh, Jay Billesberger, Matthew Brown, Daniel Cadete, Patrick Elliott, Kevin Frankowski, Kevin Head, Azad Hessamodini, Nannette Ho-Covernton, Doug Liddell, Natalia Alvarez Liebana, Tomas Malango, Warren Mitchell, Dr. John Pillay, Jim Rakievich, Hayley Sutton, Nathan Whitcombe, Heather Wilcott, and Matthew Wuthrich.

We have thoroughly enjoyed our repeat working relationship with the exceptional team of book professionals at Page Two, including Trena White, Rony Ganon, Peter Cocking, Lesley Erickson, Rachel Ironstone, Taysia Louie, Cameron McKague, Lorraine Toor, and Steph VanderMeulen.

APPENDIX:
INTERVIEW GUIDE

TO ACHIEVE some consistency across the case studies, the interviews followed a detailed interview guide across the key areas related to the acceleration of digital innovations.

- The narrative is the story that the company uses to explain why it is changing.

- Governance is the approach the company has taken to make decisions and provide oversight.

- Targeting is about how the company identifies specific areas for change.

- Support for people relates to helping employees overcome change resistance.

- The results are stories that reinforce the narrative and encourage more and faster change.

- The role of the ecosystem looks at how the company leverages others, such as startups and universities.

- The investments are specific large capital expenditures to create special-purpose facilities or funds to help boost results.

- The challenges relate to the kinds of general barriers that inhibit progress on change in the industry, such as cyber worries and capital market pressures.

The Narrative You Use

1 What is the narrative that you use to explain to your employees, stakeholders, and suppliers about the change you are seeking?

2 What are the principal drivers for your digital change journey? Survival? Growth? Brand positioning? Competitive pressures? Talent attraction? Culture? Climate change? Innovation? Capital markets? All of the above?

3 Does the narrative change depending on your business unit focus? For example, is upstream different from retail?

4 Have you articulated a vision of where your company will be in a more digital future?

Your Approach to Governance

1 How have you structured decision-making around change? A digital council? Task force? Strike team?

2 How have you organized your transformation? Is there a corporate leader, or is it decentralized to business units? Is it with a service leader (CIO, CFO)? Chief digital officer?

3 Has the organizational approach changed with time as you gain maturity?

4 Have you blended technology support teams (information technology such as ERP support, and operational technology such as SCADA support)?

How You Target Your Opportunities

1   What kinds of business issues have been targeted for digitally driven change?

2   Are change targets weighted towards revenue growth, cost reduction, asset productivity? Are they more or less balanced?

3   Have you unlocked any new creative business models that have significant impacts on the existing industry and services offered?

4   What role do customers, employees, and suppliers play in framing digital opportunity?

Your Support for Your People

1   What has been your investment in change management in supporting digital change?

2   What skills have emerged as valuable for the future?

3   How are you communicating your digital journey with your people? Which mechanisms, and what message?

4   What role does training play in preparing your people for the future?

5   Have you changed ways of working (adopting agile, for example)?

6   What is the role of the frontline worker in the digital journey? The team leader? The supervisor?

The Results You Highlight

1   What successes do you point to that communicate what you're trying to achieve?

2   Are there specific technologies that seem to be more relevant than others? Artificial intelligence? Cloud computing? Digital twin? Blockchain? Industrial Internet of Things? Augmented reality?

3   How has your view of data as an asset changed?

4   How has the pace and timing of the journey changed?

The Role of Your Ecosystem

1   How has your ecosystem of collaborators, partners, and suppliers changed in support of the journey?

2   What role, if any, has government support played in helping with the transformation?

The Investments You Make

1   Have you made any strategic investments in digital?

2   Have you set up an investment fund?

3   Have you partnered with an accelerator or innovation lab?

The Challenges You Face

1   How has the pandemic impacted your digital change journey?

2   How have commodity markets impacted your digital change journey?

3   How are your digital efforts linked to or supportive of your environmental and sustainability goals?

4   What external factors are blocking progress? Telecom network coverage?

5   How are you approaching cyber issues and pressures?

6   How are you managing ethical issues and considerations?

GLOSSARY OF TERMS

2D: two-dimensional.

3D: three-dimensional.

3D printing: a manufacturing process where an object is made by adding layers of material progressively. Also called additive manufacturing.

agile: a method of working that features quick development cycles, iterative design, multi-disciplinary teams, and minimally viable solutions.

AI: artificial intelligence.

API: application programming interface; reusable software code that allows applications to communicate with each other and exchange data, analysis, and instructions.

app: software application.

ASCII: American Standard Code for Information Interchange.

automation: a machine that replaces human labor.

autonomy: a machine that operates to a considerable degree without human supervision.

barrel: a unit of volume for crude oil, equal to 42 US gallons.

B2B: business-to-business.

beam pump: a pump used to lift hydrocarbons to the surface. Also called an iron horse.

bitumen: a very dense type of crude oil—thick and viscous.

black swan event: a rare, unpredictable set of circumstances.

blockchain: a database technology of immutable records, stored on decentralized servers.

boe: barrel of oil equivalent; a unit of energy based on the approximate energy released by combusting one barrel of crude oil. Used to combine the energy values from crude oil and natural gas into a single measure.

bot: short for robot; the outcome of using robotic process automation technologies to automate a business process.

brownfield: an existing asset or facility, usually in operations, where introducing change is time-consuming and costly.

bus: a communication system within a computer that transfers data.

business model: the design for a successful operation of a business, incorporating sources of revenue, customer description, product offering, and financing arrangements.

CAPEX: short for capital expenditure; creates assets on the balance sheet.

capital: money that is sourced from markets (stock markets, bond markets, private equity) and converted into other assets for business operations.

capital execution: spending capital on assets and construction.

carbon: shorthand for atmospheric accumulations of man-made carbon dioxide and methane emissions.

carbon neutrality: a balance between emitting carbon and removing carbon from the atmosphere.

catalyst: a chemical substance that impedes, facilitates, or accelerates a chemical reaction. Oil refining uses catalysts extensively.

change resistance: the tendency by humans to prefer the status quo over a proposed change.

cloud: data storage and computer processing services available using an internet connection.

coal measure: a coal deposit underground. Coal measures that are accessed for their gas potential and are too deep to access by mining or are too small or insufficiently rich to justify the cost of mining.

compression station: a facility for processing natural gas that compresses it into a pressurized vessel or pipeline.

compute load: the amount of computer processing power necessary for operations.

CRM: customer relationship management system; an enterprise system used to help manage a company's business interface with its customers.

crypto: shorthand for cryptocurrency; a non-sovereign store of value, such as bitcoin.

cyber: pertaining to computers; also a substitute for words that include cyber, such as cyberattack and cybersecurity.

dark web: a network of computers and servers to which one connects via special browsing software; frequently used for illicit activities such as housing stolen data or dealing in illegal trades or substances.

data lake: a super-large repository of datasets organized for common use within an organization.

data mart: a massive collection of documents and files.

DCS: decentralized control system; a computer system that is physically located close to an operating asset.

DevOps: a combination of practices and tools that allows companies to develop software products at great speed.

digital native: a person who has lived most or all of their life with digital technologies; millennials and younger.

digital twin: a virtual or software-only version of a physical asset, entire factory, or complete value chain.

downhole: refers to the insides of an oil or gas well. Instruments that measure the conditions at the bottom of a well are downhole tools.

downstream: shorthand for the distribution, wholesaling, and retailing of refined oil and gas products.

dumb metal: mechanical equipment that is not enabled with digital technology.

drone: an autonomous vehicle capable of operations with minimal human supervision.

edge computing: small self-contained computers—on machines, tools, wells, and many other assets—that provide a measure of autonomy.

edge device: a sensor, controller, or network access point that serves as an entry point for access to an enterprise.

EPC: engineering procurement construction; usually refers to an engineering company offering these services.

ERP: enterprise resource planning; a class of integrated computer software that automates a wide range of common business functions and processes.

ESG: environment, society, and governance; a management system that provides performance goals and targets beyond profitability and shareholder value.

EV: electric vehicle, whose drivetrain consists of electric motors, as compared to an internal combustion engine (ICE).

fabrication: a business process of making a unique piece of equipment or component following a design.

FEED: front end engineering and design.

fiber: telecommunications cables made of fine glass threads.

fracking: short for fracturing; refers to the injection of a mix of liquid and sand under high pressure into a rock formation, which then breaks apart (or fractures), releasing trapped hydrocarbons.

frac spread: the pumping equipment that is used in fracking.

full stack: the combined front end (user interface) and back end (database, servers, networks) of a computer system.

fungible: the ability of a good or asset to be interchanged with other individual goods or assets of the same type.

gas plant: a facility that treats raw natural gas by removing contaminants or undesirable components such as water and carbon dioxide.

gathering pipeline: a buried pipeline of (usually) narrow diameter that connects gas from a wellhead to a collection facility such as a compression station or a processing plant.

G&A: general and administrative.

GHG: greenhouse gas; usually carbon dioxide (CO_2) or methane (CH_4).

governance: the mechanisms, structures, and organization for making decisions and conducting oversight.

GPS: global positioning system; a satellite service to provide precise physical locations on Earth.

gray energy: energy derived from fossil fuels, including coal, oil, and natural gas.

green energy: energy that comes from renewable sources (solar energy, wind energy, tidal movements, flowing rivers).

greenfield: an asset or facility that exists only as a design or concept at a point at which the design is still flexible and easily changed.

heavy hauler: very large dump truck used in strip and open cut mines.

historian: the data repository of time series information collected by a supervisory control and data acquisition (SCADA) system.

hotspot: a network service on smartphones through which the phone creates a shareable internet access point.

HS2: High Speed 2; a modern high speed rail line project in the United Kingdom.

hydraulic stimulation: the technique of injecting liquid (water or chemical) into a rock formation to increase the flow of hydrocarbons to the surface.

INSEAD: Institut européen d'administration des affaires; a leading business school based in Fontainebleau, France.

IoT: Internet of Things; sensors that connect directly to the internet will dramatically outnumber phones, tablets, and other devices (the internet of websites).

IP: intellectual property.

IROC: integrated remote operations center; a control room for managing multiple remote facilities.

IT: information technology; used to reference commercial computer technology systems, such as enterprise resource planning (ERP), productivity software, email, cyber-security, and help desk services.

JOA: joint operating agreement; a contract between multiple parties involved in oil and gas production setting out their rights, obligations, and commercial terms.

KPI: key performance indicator.

lean: the practice of continuous improvement to achieve zero waste and zero defects in a process.

linear infrastructure: network assets, such as a pipeline or telecom network.

LNG: liquified natural gas; gas that is chilled to -162° Centigrade turns into a liquid, occupying 1/600 of the space required as a gas.

machine learning: a computer system that can learn and adapt without explicit instructions or changes by using algorithms and statistical models to analyze and draw inferences from patterns in data.

Metcalfe's Law: an empirical law named after Robert Metcalfe, who noted that the value of a network is related to the number of nodes on that network.

microapp: a small, proprietary application for specific technology.

midstream: shorthand for the transportation and refining sectors in oil and gas.

MOC: management of change; an engineering process to carefully introduce mechanical changes to an industrial process.

modularization: a construction technique where a large facility is built in smaller segments or modules in factory-like environments.

monetization: the technique of generating revenue from something; to monetize data means to charge for access to, or for a copy of, a dataset.

Montney: a geologic formation rich in hydrocarbons straddling the border between Northern British Columbia and Alberta.

Moore's Law: an empirical law named after Gordon Moore, who noted that the number of transistors on a microchip doubles every two years, while the cost of computers falls by 50 percent.

NLP: natural language processing; a computer technique that interprets human speech.

offshore: oil and gas operations carried out in oceans. Work carried out internationally versus domestically.

oil sands: A type of hydrocarbon deposit featuring molecules of dense oil (or bitumen) mixed with sand. Oil sands can be mined similarly to ore if near the surface.

O&M: operations and maintenance.

OPEX: operating expenditure; OPEX is reflected on the profit and loss statement.

OSDU: Open Subsurface Data Universe, a forum for promulgating open source standards for data.

OT: operational technology, includes supervisory control and data acqui-sition (SCADA) systems, process logic controllers (PLCs), and decen-tralized control systems (DCSs).

Permian Basin: a geologic formation in Texas that contains hydrocarbons.

phishing: a cyberattack where fake links or downloads deceive people into surrendering sensitive personal information.

PIG: pipeline inspection gauge. Simple PIGs de-sludge pipes, whereas intelligent PIGs measure pipe integrity.

PLC: process logic controller; a kind of computer system for controlling mechanical equipment.

pressure vessel: a tank whose contents are under pressure.

QR: quick response; a QR code is a matrix barcode invented by a Japanese automotive company, now widely used by mobile devices.

R&D: research and development.

rentier state: a state, typically commodity-based, whose income rises without an increase in national productivity.

robotics: the field of study of robots.

royalty: a contractual share of revenue from the sale of a good or service to a party otherwise uninvolved in the actual transaction.

RPA: robotic process automation; technology that automates keystrokes.

Safety Moment: a portion of a business meeting, usually at the start, to discuss safety.

SAP: a software product from the SAP software company.

SCADA: supervisory control and data acquisition; a computer system that provides real-time information about an operating asset or production process.

scrum: a technique in agile methods to coordinate activities of workers.

seismic: relating to geologic vibrations such as earth tremors or earthquakes.

separation unit: a facility for processing hydrocarbons that separates out various valuable compounds, usually by evaporation and distillation.

shut-down: a practice in heavy industry where energized plants (under pressure, heat, or high power) are slowly de-energized to enable maintenance activities on the plant.

Slack: a cloud-based collaboration platform widely used in the software development industry.

smart: the presence of digital capabilities within a traditionally non-digital thing.

smart contract: a feature of blockchain databases whereby actions agreed in a contract (such as payment on a certain date) are automatically executed by the blockchain system.

sour: oil or gas with high sulfur content.

spoofing: a type of cyberattack in which a malicious link is disguised as a safe one.

stack construction: a construction acceleration technique for buildings where segments of a floor are built in a factory and assembled on the construction site.

stand-up: a meeting during which the participants stand throughout to ensure a faster pace.

stiction: the degradation of coatings on a magnetic surface, rendering the magnetic surface unreadable.

straddle plant: a kind of gas plant that removes natural gas liquids from a gas stream.

subsurface data: data about rock formations underground.

SVP: senior vice president

sweet: oil or gas with low sulfur content.

Sword of Damocles: a situation of impending doom.

take-or-pay: a contracting model in which the charge for usage does not vary with volume.

telework(ing): work that is carried out remotely via telecommunications access to company computer systems.

Technology Life Cycle: coined by Nikolai Kondratiev; describes the process by which a novel technology evolves from research and development (R&D) to saturation.

TIBCO Spotfire: data visualization software by The Information Bus Company.

token: a record on a blockchain database.

tokenizing, tokenization: the use of an immutable record on a blockchain database to record ownership, an event, or an identity.

track and trace: a technique used to monitor raw materials from source through industrial processes and final consumption.

tubular: hollow, seamless cylindrical tubes made of steel, used throughout the oil and gas industry; pipe.

turnaround: a technique in heavy industry to shut down an energized asset and carry out configuration changes to change the mix of produced products.

turnkey: a term used to describe an asset that on delivery to a customer requires minimal, if any, additional activities by the customer before the asset can be used.

UAV: unmanned aerial vehicle, or flying drone.

unitization: building components that are shop-prepared, assembled, and shipped to the work site as a completed unit.

upstream: shorthand for oil and gas exploration and production.

USD: US dollar; the currency of the United States.

UX: user experience. The interface between a person and a digital technology. Well-designed interfaces are easy to use, require no training, and lead to a positive user experience.

VAKT: a software company.

virtualization: the technique of creating a virtual or software-only version of a physical thing.

visual analytics: a field of artificial intelligence and machine learning through which algorithms analyze data collected by a visual sensor, interpret the data, and enable an automated action to happen.

visualization: the technique of creating a chart or other image of a set of information, an object, or a situation.

well pad: a cleared area of land where an oil or gas well is drilled.

well log: a record of the various geologic formations and layers penetrated by a borehole.

Wi-Fi: wireless fidelity; a mechanism that allows mobile devices to connect wirelessly to each other or the internet within an area.

wiki: a publication that is collaboratively edited and managed by its own audience.

Yammer: a collaboration software product from Microsoft.

INDEX

240–45; Worley case study, 245–49.
See also engineering, procurement,
and construction (EPC) companies
shale oil, American, 15
shared assets, 134–35
SharePoint by Microsoft, 214
sharing economy, 29
Shell, 1, 12, 26, 121, 149, 229, 230
ships and vessels, 73, 165, 247
short countries, 19
shovel, mine, 208
signature digital events, 253
smart contracts, 83, 223
smartphones, 56, 67, 78
smart things, 91. *See also* edge devices;
industrial Internet of Things; sensors
SoftBank, 136
software developers, 210
sponsorship, for digital change, 181, 252
stand-up meetings, 55
subsurface data, 19
success, setting up for, 255
Suncor, 164
suppliers, 18, 258. *See also* equipment sector
supply chains: bulk commodity transfers, 81; digital reconfiguration, 43;
digital twins and, 139–40; lags in
digital innovation, 20; robots and,
76, 122; tracing, 125–30, 131–32;
transparency, 43; trust and verifica-
tion issues, 106–7
support: for digital change, 181, 252;
NorthRiver Midstream and, 227–28;
for talent, 155, 257–58; for technology, 190
surplus and surplus goods, 88–89
surveillance, 122
surveys, for digital readiness, 154–55
sustainability, 29. *See also* European
Green Deal
system architects, 210

T

talent, 182–93; accepting the limitations of, 220–21; assembling new
skills, 257–58; building new skills,
258; change resistance and recruit-
ment challenges, 18, 21, 160;
changers vs. workers, 152; changing
roles in oil and gas industry, 103–4;
concerns among about future,
182–84; as consultants and contractors, 192–93; creating a career,
190–91; digital leadership pipeline
challenges, 175–77; engagement
with during digital change, 204–5;
entrepreneurship, 192; framing
the narrative for, 251; helping
with career transitions, 188–89;
high-demand jobs, 189–90; human
resources and robotic process automation (RPA), 75–76; Imperial Oil
and, 209–10; innovation teams,
232–33; leveraging emotions
of, 184–86; location agnosticism
and, 256; managing transitions,
228–29; mass customization of
job roles, 234; NAL Resources and,
220–21; NorthRiver Midstream and,
225, 228–29; redesigning work for
autonomy vs. humans, 103–4, 121,
165–66, 264–65; selection criteria,
257; sourcing models, 257; stages
of digital adoption by, 186–88, *187*;
support for, 155, 257–58; teamwork, 104–5; during unemployment
periods, 191–93; VARO Energy
and, 232–33; Wood and, 241;
Worley and, 248. *See also* change
resistance; digital leadership;
people-lite businesses
taxi industry, 98, 100, *101*, 238
teamwork, 104–5. *See also*
collaboration

ABOUT THE AUTHORS

GEOFFREY CANN is an author, professional speaker, and trainer with a mission to help the oil and gas industry embrace digital innovation as a means to improve its environmental and economic performance, and to buy time for a successful energy transition. Following an early career with Imperial Oil, he joined Deloitte and worked as a management consultant to companies around the world. Today, he specializes in digital innovation in the energy sector, writes and publishes regularly on this topic, and teaches an executive course on digital awareness. He published his first book, *Bits, Bytes, and Barrels: The Digital Transformation of Oil and Gas*, in January 2019, with co-author Rachael Goydan. He is also a contributing author for the book *Machine Learning and Data Science in the Oil and Gas Industry: Best Practices, Tools, and Case Studies*, published in March 2021. He is an alumnus of McGill University (BComm, 1984) and the Ivey Business School at the University of Western Ontario (MBA, 1989).

RYAN CANN is a researcher and analyst. With both a bachelor's and a master's degree in English, he focuses his writing and research skills on identifying where digital tools such as artificial intelligence, blockchain, cloud computing, automation, and other technologies are having impacts on the energy sector. His focus is primarily in oil and gas, but he has also investigated manufacturing, medicine, automotive, and agriculture sectors. He is an alumnus of Ryerson University (BA, 2017) and the University of Toronto (MA, 2018).